THE POWER OF HABIT

THE POWER OF

HABIT

Why We Do What We Do
and How to Change

CHARLES DUHIGG

WILLIAM HEINEMANN: LONDON

Published by William Heinemann 2012

4 6 8 10 9 7 5

Copyright © Charles Duhigg 2012

Charles Duhigg has asserted his right under the Copyright, Designs
and Patents Act, 1988, to be identified as the author of this work.

Illustration on page 189 by Andrew Pole
All other illustrations by Anton Ioukhnovets

First published in Great Britain in 2012 by
William Heinemann
Random House, 20 Vauxhall Bridge Road,
London SW1V 2SA

www.randomhouse.co.uk

Addresses for companies within The Random House Group Limited can be found at:
www.randomhouse.co.uk/offices.htm

The Random House Group Limited Reg. No. 954009

A CIP catalogue record for this book is available from the British Library

ISBN 9780434020362

The Random House Group Limited supports The Forest Stewardship
Council (FSC®), the leading international forest certification organisation. Our books
carrying the FSC label are printed on FSC® certified paper. FSC is the only forest
certification scheme endorsed by the leading environmental organisations,
including Greenpeace. Our paper procurement policy can be found at:
www.randomhouse.co.uk/environment

Printed and bound by CPI Group (UK) Ltd, Croydon, CR0 4YY

To Oliver, John Harry,
John and Doris,
and, everlastingly, to Liz

CONTENTS

● ● ●

PART THREE
The Habits of Societies

● ● ●

THE POWER OF HABIT

PROLOGUE

The Habit Cure

She was the scientists' favorite participant.

Lisa Allen, according to her file, was thirty-four years old, had started smoking and drinking when she was sixteen, and had struggled with obesity for most of her life. At one point, in her mid-twenties, collection agencies were hounding her to recover $10,000 in debts. An old résumé listed her longest job as lasting less than a year.

The woman in front of the researchers today, however, was lean and vibrant, with the toned legs of a runner. She looked a decade younger than the photos in her chart and like she could out-exercise anyone in the room. According to the most recent report in her file, Lisa had no outstanding debts, didn't drink, and was in her thirty-ninth month at a graphic design firm.

"How long since your last cigarette?" one of the physicians asked, starting down the list of questions Lisa answered every time she came to this laboratory outside Bethesda, Maryland.

"Almost four years," she said, "and I've lost sixty pounds and run a marathon since then." She'd also started a master's degree and bought a home. It had been an eventful stretch.

The scientists in the room included neurologists, psychologists, geneticists, and a sociologist. For the past three years, with funding from the National Institutes of Health, they had poked and prodded Lisa and more than two dozen other former smokers, chronic over-eaters, problem drinkers, obsessive shoppers, and people with other destructive habits. All of the participants had one thing in common: They had remade their lives in relatively short periods of time. The researchers wanted to understand how. So they measured subjects' vital signs, installed video cameras inside their homes to watch their daily routines, sequenced portions of their DNA, and, with tech-nologies that allowed them to peer inside people's skulls in real time, watched as blood and electrical impulses flowed through their brains while they were exposed to temptations such as cigarette smoke and lavish meals. The researchers' goal was to figure out how habits work on a neurological level—and what it took to make them change.

"I know you've told this story a dozen times," the doctor said to Lisa, "but some of my colleagues have only heard it secondhand. Would you mind describing again how you gave up cigarettes?"

"Sure," Lisa said. "It started in Cairo." The vacation had been something of a rash decision, she explained. A few months earlier, her husband had come home from work and announced that he was leaving her because he was in love with another woman. It took Lisa a while to process the betrayal and absorb the fact that she was actually getting a divorce. There was a period of mourning, then a period of obsessively spying on him, following his new girlfriend around town, calling her after midnight and hanging up. Then there was the evening Lisa showed up at the girlfriend's house, drunk, pounding on her door and screaming that she was going to burn the condo down.

"It wasn't a great time for me," Lisa said. "I had always wanted to see the pyramids, and my credit cards weren't maxed out yet, so . . ."

On her first morning in Cairo, Lisa woke at dawn to the sound of the call to prayer from a nearby mosque. It was pitch black inside her hotel room. Half blind and jet-lagged, she reached for a cigarette.

She was so disoriented that she didn't realize—until she smelled burning plastic—that she was trying to light a pen, not a Marlboro. She had spent the past four months crying, binge eating, unable to sleep, and feeling ashamed, helpless, depressed, and angry, all at once. Lying in bed, she broke down. "It was like this wave of sadness," she said. "I felt like everything I had ever wanted had crumbled. I couldn't even smoke right.

"And then I started thinking about my ex-husband, and how hard it would be to find another job when I got back, and how much I was going to hate it and how unhealthy I felt all the time. I got up and knocked over a water jug and it shattered on the floor, and I started crying even harder. I felt desperate, like I had to change something, at least one thing I could control."

She showered and left the hotel. As she rode through Cairo's rutted streets in a taxi and then onto the dirt roads leading to the Sphinx, the pyramids of Giza, and the vast, endless desert around them, her self-pity, for a brief moment, gave way. She needed a goal in her life, she thought. Something to work toward.

So she decided, sitting in the taxi, that she would come back to Egypt and trek through the desert.

It was a crazy idea, Lisa knew. She was out of shape, overweight, with no money in the bank. She didn't know the name of the desert she was looking at or if such a trip was possible. None of that mattered, though. She needed something to focus on. Lisa decided that she would give herself one year to prepare. And to survive such an expedition, she was certain she would have to make sacrifices.

In particular, she would need to quit smoking.

When Lisa finally made her way across the desert eleven months later—in an air-conditioned and motorized tour with a half-dozen other people, mind you—the caravan carried so much water, food, tents, maps, global positioning systems, and two-way radios that throwing in a carton of cigarettes wouldn't have made much of a difference.

But in the taxi, Lisa didn't know that. And to the scientists at the laboratory, the details of her trek weren't relevant. Because for reasons they were just beginning to understand, that one small shift in Lisa's perception that day in Cairo—the conviction that she *had* to give up smoking to accomplish her goal—had touched off a series of changes that would ultimately radiate out to every part of her life. Over the next six months, she would replace smoking with jogging, and that, in turn, changed how she ate, worked, slept, saved money, scheduled her workdays, planned for the future, and so on. She would start running half-marathons, and then a marathon, go back to school, buy a house, and get engaged. Eventually she was recruited into the scientists' study, and when researchers began examining images of Lisa's brain, they saw something remarkable: One set of neurological patterns—her old habits—had been overridden by new patterns. They could still see the neural activity of her old behaviors, but those impulses were crowded out by new urges. As Lisa's habits changed, so had her brain.

It wasn't the trip to Cairo that had caused the shift, scientists were convinced, or the divorce or desert trek. It was that Lisa had focused on changing just one habit—smoking—at first. Everyone in the study had gone through a similar process. By focusing on one pattern—what is known as a "keystone habit"—Lisa had taught herself how to reprogram the other routines in her life, as well.

It's not just individuals who are capable of such shifts. When companies focus on changing habits, whole organizations can transform. Firms such as Procter & Gamble, Starbucks, Alcoa, and Target have seized on this insight to influence how work gets done, how employees communicate, and—without customers realizing it—the way people shop.

"I want to show you one of your most recent scans," a researcher told Lisa near the end of her exam. He pulled up a picture on a computer screen that showed images from inside her head. "When you see food, these areas"—he pointed to a place near the center of her brain—"which are associated with craving and hunger, are still active. Your brain still produces the urges that made you overeat.

"However, there's new activity in this area"—he pointed to the region closest to her forehead—"where we believe behavioral inhibition and self-discipline starts. That activity has become more pronounced each time you've come in."

Lisa was the scientists' favorite participant because her brain scans were so compelling, so useful in creating a map of where behavioral patterns—habits—reside within our minds. "You're helping us understand how a decision becomes an automatic behavior," the doctor told her.

Everyone in the room felt like they were on the brink of something important. And they were.

● ● ●

When you woke up this morning, what did you do first? Did you hop in the shower, check your email, or grab a doughnut from the kitchen counter? Did you brush your teeth before or after you toweled off? Tie the left or right shoe first? What did you say to your kids on your way out the door? Which route did you drive to work? When you got to your desk, did you deal with email, chat with a colleague, or jump into writing a memo? Salad or hamburger for lunch? When you got home, did you put on your sneakers and go for a run, or pour yourself a drink and eat dinner in front of the TV?

"All our life, so far as it has definite form, is but a mass of habits," William James wrote in 1892. Most of the choices we make each day may feel like the products of well-considered decision making, but they're not. They're habits. And though each habit means relatively little on its own, over time, the meals we order, what we say to our kids each night, whether we save or spend, how often we exercise, and the way we organize our thoughts and work routines have enormous impacts on our health, productivity, financial security, and happiness. One paper published by a Duke University researcher in 2006 found that more than 40 percent of the actions people performed each day weren't actual decisions, but habits.

William James—like countless others, from Aristotle to Oprah—spent much of his life trying to understand why habits exist. But only in the past two decades have scientists and marketers really begun understanding how habits *work*—and more important, how they change.

This book is divided into three parts. The first section focuses on how habits emerge within individual lives. It explores the neurology of habit formation, how to build new habits and change old ones, and the methods, for instance, that one ad man used to push toothbrushing from an obscure practice into a national obsession. It shows how Procter & Gamble turned a spray named Febreze into a billion-dollar business by taking advantage of consumers' habitual urges, how Alcoholics Anonymous reforms lives by attacking habits at the core of addiction, and how coach Tony Dungy reversed the fortunes of the worst team in the National Football League by focusing on his players' automatic reactions to subtle on-field cues.

The second part examines the habits of successful companies and organizations. It details how an executive named Paul O'Neill—before he became treasury secretary—remade a struggling aluminum manufacturer into the top performer in the Dow Jones Industrial Average by focusing on one keystone habit, and how Starbucks turned a high school dropout into a top manager by instilling habits designed to strengthen his willpower. It describes why even the most talented surgeons can make catastrophic mistakes when a hospital's organizational habits go awry.

The third part looks at the habits of societies. It recounts how Martin Luther King, Jr., and the civil rights movement succeeded, in part, by changing the ingrained social habits of Montgomery, Alabama—and why a similar focus helped a young pastor named Rick Warren build the nation's largest church in Saddleback Valley, California. Finally, it explores thorny ethical questions, such as whether a murderer in Britain should go free if he can convincingly argue that his habits led him to kill.

Each chapter revolves around a central argument: Habits can be changed, if we understand how they work.

This book draws on hundreds of academic studies, interviews with more than three hundred scientists and executives, and research conducted at dozens of companies. (For an index of resources, please see the book's notes and http://www.thepowerof habit.com.) It focuses on habits as they are technically defined: the choices that all of us deliberately make at some point, and then stop thinking about but continue doing, often every day. At one point, we all consciously decided how much to eat and what to focus on when we got to the office, how often to have a drink or when to go for a jog. Then we stopped making a choice, and the behavior became automatic. It's a natural consequence of our neurology. And by understanding how it happens, you can rebuild those patterns in whichever way you choose.

● ● ●

I first became interested in the science of habits eight years ago, as a newspaper reporter in Baghdad. The U.S. military, it occurred to me as I watched it in action, is one of the biggest habit-formation experiments in history. Basic training teaches soldiers carefully designed habits for how to shoot, think, and communicate under fire. On the battlefield, every command that's issued draws on behaviors practiced to the point of automation. The entire organization relies on endlessly rehearsed routines for building bases, setting strategic priorities, and deciding how to respond to attacks. In those early days of the war, when the insurgency was spreading and death tolls were mounting, commanders were looking for habits they could instill among soldiers and Iraqis that might create a durable peace.

I had been in Iraq for about two months when I heard about an officer conducting an impromptu habit modification program in Kufa, a small city ninety miles south of the capital. He was an army major who had analyzed videotapes of recent riots and had identified a pattern: Violence was usually preceded by a crowd of Iraqis gathering in a plaza or other open space and, over the course of several hours, growing in size. Food vendors would show up, as well as spectators. Then, someone would throw a rock or a bottle and all hell would break loose.

When the major met with Kufa's mayor, he made an odd request: Could they keep food vendors out of the plazas? Sure, the mayor said. A few weeks later, a small crowd gathered near the Masjid al-Kufa, or Great Mosque of Kufa. Throughout the afternoon, it grew in size. Some people started chanting angry slogans. Iraqi police, sensing trouble, radioed the base and asked U.S. troops to stand by. At dusk, the crowd started getting restless and hungry. People looked for the kebab sellers normally filling the plaza, but there were none to be found. The spectators left. The chanters became dispirited. By 8 P.M., everyone was gone.

When I visited the base near Kufa, I talked to the major. You wouldn't necessarily think about a crowd's dynamics in terms of habits, he told me. But he had spent his entire career getting drilled in the psychology of habit formation.

At boot camp, he had absorbed habits for loading his weapon, falling asleep in a war zone, maintaining focus amid the chaos of battle, and making decisions while exhausted and overwhelmed. He had attended classes that taught him habits for saving money, exercising each day, and communicating with bunkmates. As he moved up the ranks, he learned the importance of organizational habits in ensuring that subordinates could make decisions without constantly asking permission, and how the right routines made it easier to work alongside people he normally couldn't stand. And now, as an impromptu nation builder, he was seeing how crowds and cultures abided by many of the same rules. In some sense, he said, a community was a giant collection of habits occurring among thousands of people that, depending on how they're influenced, could result in violence or peace. In addition to removing the food vendors, he had launched dozens of different experiments in Kufa to influence residents' habits. There hadn't been a riot since he arrived.

"Understanding habits is the most important thing I've learned in the army," the major told me. "It's changed everything about how I see the world. You want to fall asleep fast and wake up feeling good? Pay attention to your nighttime patterns and what you automatically do when you get up. You want to make running easy? Create triggers to make it a routine. I drill my kids on this stuff. My wife and I write out habit plans for our marriage. This is all we talk about in command meetings. Not one person in Kufa would have told me that we could influence crowds by taking away the kebab stands, but once you see everything as a bunch of habits, it's like someone gave you a flashlight and a crowbar and you can get to work."

The major was a small man from Georgia. He was perpetually spitting either sunflower seeds or chewing tobacco into a cup. He told me that prior to entering the military, his best career option had been repairing telephone lines, or, possibly, becoming a methamphetamine entrepreneur, a path some of his high school peers had chosen to less success. Now, he oversaw eight hundred troops in one of the most sophisticated fighting organizations on earth.

"I'm telling you, if a hick like me can learn this stuff, anyone can. I tell my soldiers all the time, there's nothing you can't do if you get the habits right."

In the past decade, our understanding of the neurology and psychology of habits and the way patterns work within our lives, societies, and organizations has expanded in ways we couldn't have imagined fifty years ago. We now know why habits emerge, how they change, and the science behind their mechanics. We know how to break them into parts and rebuild them to our specifications. We understand how to make people eat less, exercise more, work more efficiently, and live healthier lives. Transforming a habit isn't necessarily easy or quick. It isn't always simple.

But it is possible. And now we understand how.

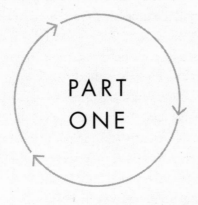

PART ONE

The Habits of Individuals

1

THE HABIT LOOP
How Habits Work

I.

In the fall of 1993, a man who would upend much of what we know about habits walked into a laboratory in San Diego for a scheduled appointment. He was elderly, a shade over six feet tall, and neatly dressed in a blue button-down shirt. His thick white hair would have inspired envy at any fiftieth high school reunion. Arthritis caused him to limp slightly as he paced the laboratory's hallways, and he held his wife's hand, walking slowly, as if unsure about what each new step would bring.

About a year earlier, Eugene Pauly, or "E.P." as he would come to be known in medical literature, had been at home in Playa del Rey, preparing for dinner, when his wife mentioned that their son, Michael, was coming over.

"Who's Michael?" Eugene asked.

"Your child," said his wife, Beverly. "You know, the one we raised?"

Eugene looked at her blankly. "Who is that?" he asked.

The next day, Eugene started vomiting and writhing with stomach cramps. Within twenty-four hours, his dehydration was so pronounced that a panicked Beverly took him to the emergency room. His temperature started rising, hitting 105 degrees as he sweated a yellow halo of perspiration onto the hospital's sheets. He became delirious, then violent, yelling and pushing when nurses tried to insert an IV into his arm. Only after sedation was a physician able to slide a long needle between two vertebra in the small of his back and extract a few drops of cerebrospinal fluid.

The doctor performing the procedure sensed trouble immediately. The fluid surrounding the brain and spinal nerves is a barrier against infection and injury. In healthy individuals, it is clear and quick flowing, moving with an almost silky rush through a needle. The sample from Eugene's spine was cloudy and dripped out sluggishly, as if filled with microscopic grit. When the results came back from the laboratory, Eugene's physicians learned why he was ill: He was suffering from viral encephalitis, a disease caused by a relatively harmless virus that produces cold sores, fever blisters, and mild infections on the skin. In rare cases, however, the virus can make its way into the brain, inflicting catastrophic damage as it chews through the delicate folds of tissue where our thoughts, dreams—and according to some, souls—reside.

Eugene's doctors told Beverly there was nothing they could do to counter the damage already done, but a large dose of antiviral drugs might prevent it from spreading. Eugene slipped into a coma and for ten days was close to death. Gradually, as the drugs fought the disease, his fever receded and the virus disappeared. When he finally awoke, he was weak and disoriented and couldn't swallow properly. He couldn't form sentences and would sometimes gasp, as if he had momentarily forgotten how to breathe. But he was alive.

Eventually, Eugene was well enough for a battery of tests. The doctors were amazed to find that his body—including his nervous system—appeared largely unscathed. He could move his limbs and was responsive to noise and light. Scans of his head, though, revealed ominous shadows near the center of his brain. The virus had destroyed an oval of tissue close to where his cranium and spinal column met. "He might not be the person you remember," one doctor warned Beverly. "You need to be ready if your husband is gone."

Eugene was moved to a different wing of the hospital. Within a week, he was swallowing easily. Another week, and he started talking normally, asking for Jell-O and salt, flipping through television channels and complaining about boring soap operas. By the time he was discharged to a rehabilitation center five weeks later, Eugene was walking down hallways and offering nurses unsolicited advice about their weekend plans.

"I don't think I've ever seen anyone come back like this," a doctor told Beverly. "I don't want to raise your hopes, but this is amazing."

Beverly, however, remained concerned. In the rehab hospital it became clear that the disease had changed her husband in unsettling ways. Eugene couldn't remember which day of the week it was, for instance, or the names of his doctors and nurses, no matter how many times they introduced themselves. "Why do they keep asking me all these questions?" he asked Beverly one day after a physician left his room. When he finally returned home, things got even stranger. Eugene didn't seem to remember their friends. He had trouble following conversations. Some mornings, he would get out of bed, walk into the kitchen, cook himself bacon and eggs, then climb back under the covers and turn on the radio. Forty minutes later, he would do the same thing: get up, cook bacon and eggs, climb back into bed, and fiddle with the radio. Then he would do it again.

Alarmed, Beverly reached out to specialists, including a researcher at the University of California, San Diego, who specialized in memory loss. Which is how, on a sunny fall day, Beverly and Eugene found themselves in a nondescript building on the university's campus, holding hands as they walked slowly down a hallway. They were shown into a small exam room. Eugene began chatting with a young woman who was using a computer.

"Having been in electronics over the years, I'm amazed at all this," he said, gesturing at the machine she was typing on. "When I was younger, that thing would have been in a couple of six-foot racks and taken up this whole room."

The woman continued pecking at the keyboard. Eugene chuckled.

"That is incredible," he said. "All those printed circuits and diodes and triodes. When I was in electronics, there would have been a couple of six-foot racks holding that thing."

A scientist entered the room and introduced himself. He asked Eugene how old he was.

"Oh, let's see, fifty-nine or sixty?" Eugene replied. He was seventy-one years old.

The scientist started typing on the computer. Eugene smiled and pointed at it. "That is really something," he said. "You know, when I was in electronics there would have been a couple of six-foot racks holding that thing!"

The scientist was fifty-two-year-old Larry Squire, a professor who had spent the past three decades studying the neuroanatomy of memory. His specialty was exploring how the brain stores events. His work with Eugene, however, would soon open a new world to him and hundreds of other researchers who have reshaped our understanding of how habits function. Squire's studies would show that even someone who can't remember his own age or almost anything else can develop habits that seem inconceivably complex—until you realize that everyone relies on similar neurological processes every day. His and others' research would help reveal the subconscious mechanisms that impact the countless choices that seem as if they're the products of well-reasoned thought, but actually are in-fluenced by urges most of us barely recognize or understand.

By the time Squire met Eugene, he had already been studying images of his brain for weeks. The scans indicated that almost all the damage within Eugene's skull was limited to a five-centimeter area near the center of his head. The virus had almost entirely de-stroyed his medial temporal lobe, a sliver of cells which scientists suspected was responsible for all sorts of cognitive tasks such as recall of the past and the regulation of some emotions. The com-pleteness of the destruction didn't surprise Squire—viral encephali-tis consumes tissue with a ruthless, almost surgical, precision. What shocked him was how familiar the images seemed.

Thirty years earlier, as a PhD student at MIT, Squire had worked alongside a group studying a man known as "H.M.," one of the most famous patients in medical history. When H.M.—his real name was Henry Molaison, but scientists shrouded his identity throughout his life—was seven years old, he was hit by a bicycle and landed hard on his head. Soon afterward, he developed seizures and started blacking out. At sixteen, he had his first grand mal seizure, the kind that affects the entire brain; soon, he was losing conscious-ness up to ten times a day.

By the time he turned twenty-seven, H.M. was desperate. Anti-convulsive drugs hadn't helped. He was smart, but couldn't hold a job. He still lived with his parents. H.M. wanted a normal existence. So he sought help from a physician whose tolerance for experimen-tation outweighed his fear of malpractice. Studies had suggested that an area of the brain called the hippocampus might play a role in seizures. When the doctor proposed cutting into H.M.'s head, lift-ing up the front portion of his brain, and, with a small straw, suck-ing out the hippocampus and some surrounding tissue from the interior of his skull, H.M. gave his consent.

The surgery occurred in 1953, and as H.M. healed, his seizures slowed. Almost immediately, however, it became clear that his brain had been radically altered. H.M. knew his name and that his mother was from Ireland. He could remember the 1929 stock market crash and news reports about the invasion of Normandy. But almost everything that came afterward—all the memories, experiences, and struggles from most of the decade before his surgery—had been erased. When a doctor began testing H.M.'s memory by showing him playing cards and lists of numbers, he discovered that H.M. couldn't retain any new information for more than twenty seconds or so.

From the day of his surgery until his death in 2008, every person H.M. met, every song he heard, every room he entered, was a completely fresh experience. His brain was frozen in time. Each day, he was befuddled by the fact that someone could change the television channel by pointing a black rectangle of plastic at the screen. He introduced himself to his doctors and nurses over and over, dozens of times each day.

"I loved learning about H.M., because memory seemed like such a tangible, exciting way to study the brain," Squire told me. "I grew up in Ohio, and I can remember, in first grade, my teacher handing everyone crayons, and I started mixing all the colors together to see if it would make black. Why have I kept that memory, but I can't remember what my teacher looked like? Why does my brain decide that one memory is more important than another?"

When Squire received the images of Eugene's brain, he marveled at how similar it seemed to H.M.'s. There were empty, walnut-sized chunks in the middle of both their heads. Eugene's memory—just like H.M.'s—had been removed.

As Squire began examining Eugene, though, he saw that this patient was different from H.M. in some profound ways. Whereas almost everyone knew within minutes of meeting H.M. that something was amiss, Eugene could carry on conversations and perform tasks that wouldn't alert a casual observer that anything was wrong.

The effects of H.M.'s surgery had been so debilitating that he was institutionalized for the remainder of his life. Eugene, on the other hand, lived at home with his wife. H.M. couldn't really carry on conversations. Eugene, in contrast, had an amazing knack for guiding almost any discussion to a topic he was comfortable talking about at length, such as satellites—he had worked as a technician for an aerospace company—or the weather.

Squire started his exam of Eugene by asking him about his youth. Eugene talked about the town where he had grown up in central California, his time in the merchant marines, a trip he had taken to Australia as a young man. He could remember most of the events in his life that had occurred prior to about 1960. When Squire asked about later decades, Eugene politely changed the topic and said he had trouble recollecting some recent events.

Squire conducted a few intelligence tests and found that Eugene's intellect was still sharp for a man who couldn't remember the last three decades. What's more, Eugene still had all the habits he had formed in his youth, so whenever Squire gave him a cup of water or complimented him on a particularly detailed answer, Eugene would thank him and offer a compliment in return. Whenever someone entered the room, Eugene would introduce himself and ask about their day.

But when Squire asked Eugene to memorize a string of numbers or describe the hallway outside the laboratory's door, the doctor found his patient couldn't retain any new information for more than a minute or so. When someone showed Eugene photos of his grandchildren, he had no idea who they were. When Squire asked if he remembered getting sick, Eugene said he had no recollection of his illness or the hospital stay. In fact, Eugene almost never recalled that he was suffering from amnesia. His mental image of himself didn't include memory loss, and since he couldn't remember the injury, he couldn't conceive of anything being wrong.

In the months after meeting Eugene, Squire conducted experiments that tested the limits of his memory. By then, Eugene and Beverly had moved from Playa del Rey to San Diego to be closer to their daughter, and Squire often visited their home for his exams. One day, Squire asked Eugene to sketch a layout of his house. Eugene couldn't draw a rudimentary map showing where the kitchen or bedroom was located. "When you get out of bed in the morning, how do you leave your room?" Squire asked.

"You know," Eugene said, "I'm not really sure."

Squire took notes on his laptop, and as the scientist typed, Eugene became distracted. He glanced across the room and then stood up, walked into a hallway, and opened the door to the bathroom. A few minutes later, the toilet flushed, the faucet ran, and Eugene, wiping his hands on his pants, walked back into the living room and sat down again in his chair next to Squire. He waited patiently for the next question.

At the time, no one wondered how a man who couldn't draw a map of his home was able to find the bathroom without hesitation. But that question, and others like it, would eventually lead to a trail of discoveries that has transformed our understanding of habits' power. It would help spark a scientific revolution that today involves hundreds of researchers who are learning, for the first time, to understand all the habits that influence our lives.

As Eugene sat at the table, he looked at Squire's laptop.

"That's amazing," he said, gesturing at the computer. "You know, when I was in electronics, there would have been a couple of six-foot racks holding that thing."

● ● ●

In the first few weeks after they moved into their new house, Beverly tried to take Eugene outside each day. The doctors had told her that it was important for him to get exercise, and if Eugene was inside too long he drove Beverly crazy, asking her the same questions over and over in an endless loop. So each morning and afternoon, she took him on a walk around the block, always together and always along the same route.

The doctors had warned Beverly that she would need to monitor Eugene constantly. If he ever got lost, they said, he would never be able to find his way home. But one morning, while she was getting dressed, Eugene slipped out the front door. He had a tendency to wander from room to room, so it took her a while to notice he was gone. When she did, she became frantic. She ran outside and scanned the street. She couldn't see him. She went to the neighbors' house and pounded on the windows. Their homes looked similar— maybe Eugene had become confused and had gone inside? She ran to the door and rang the bell until someone answered. Eugene wasn't there. She sprinted back to the street, running up the block, screaming Eugene's name. She was crying. What if he had wandered into traffic? How would he tell anyone where he lived? She had been outside for fifteen minutes already, looking everywhere. She ran home to call the police.

When she burst through the door, she found Eugene in the living room, sitting in front of the television watching the History Channel. Her tears confused him. He didn't remember leaving, he said, didn't know where he'd been, and couldn't understand why she was so upset. Then Beverly saw a pile of pinecones on the table, like the ones she'd seen in a neighbor's yard down the street. She came closer and looked at Eugene's hands. His fingers were sticky with sap. That's when she realized that Eugene had gone for a walk by himself. He had wandered down the street and collected some souvenirs.

And he had found his way home.

Soon, Eugene was going for walks every morning. Beverly tried to stop him, but it was pointless.

"Even if I told him to stay inside, he wouldn't remember a few minutes later," she told me. "I followed him a few times to make sure he wouldn't get lost, but he always came back." Sometimes he would return with pinecones or rocks. Once he came back with a wallet; another time with a puppy. He never remembered where they came from.

When Squire and his assistants heard about these walks, they started to suspect that something was happening inside Eugene's head that didn't have anything to do with his conscious memory. They designed an experiment. One of Squire's assistants visited the house one day and asked Eugene to draw a map of the block where he lived. He couldn't do it. How about where his house was located on the street, she asked. He doodled a bit, then forgot the assignment. She asked him to point out which doorway led to the kitchen. Eugene looked around the room. He didn't know, he said. She asked Eugene what he would do if he were hungry. He stood up, walked into the kitchen, opened a cabinet, and took down a jar of nuts.

Later that week, a visitor joined Eugene on his daily stroll. They walked for about fifteen minutes through the perpetual spring of Southern California, the scent of bougainvillea heavy in the air. Eugene didn't say much, but he always led the way and seemed to know where he was going. He never asked for directions. As they rounded the corner near his house, the visitor asked Eugene where he lived. "I don't know, exactly," he said. Then he walked up his sidewalk, opened his front door, went into the living room, and turned on the television.

It was clear to Squire that Eugene was absorbing new information. But where inside his brain was that information residing? How could someone find a jar of nuts when he couldn't say where the kitchen was located? Or find his way home when he had no idea which house was his? How, Squire wondered, were new patterns forming inside Eugene's damaged brain?

II.

Within the building that houses the Brain and Cognitive Sciences department of the Massachusetts Institute of Technology are laboratories that contain what, to the casual observer, look like dollhouse versions of surgical theaters. There are tiny scalpels, small drills, and miniature saws less than a quarter inch wide attached to robotic arms. Even the operating tables are tiny, as if prepared for child-sized surgeons. The rooms are always kept at a chilly sixty degrees because a slight nip in the air steadies researchers' fingers during delicate procedures. Inside these laboratories, neurologists cut into the skulls of anesthetized rats, implanting tiny sensors that can record the smallest changes inside their brains. When the rats wake, they hardly seem to notice that there are now dozens of microscopic wires arrayed, like neurological spider webs, inside their heads.

These laboratories have become the epicenter for a quiet revolution in the science of habit formation, and the experiments unfolding here explain how Eugene—as well as you, me, and everyone else—developed the behaviors necessary to make it through each day. The rats in these labs have illuminated the complexity that occurs inside our heads whenever we do something as mundane as brush our teeth or back the car out of the driveway. And for Squire, these laboratories helped explain how Eugene managed to learn new habits.

When the MIT researchers started working on habits in the 1990s—at about the same time that Eugene came down with his fever—they were curious about a nub of neurological tissue known as the basal ganglia. If you picture the human brain as an onion, composed of layer upon layer of cells, then the outside layers—those closest to the scalp—are generally the most recent additions from an evolutionary perspective. When you dream up a new invention or laugh at a friend's joke, it's the outside parts of your brain at work. That's where the most complex thinking occurs.

Deeper inside the brain and closer to the brain stem—where the brain meets the spinal column—are older, more primitive structures. They control our automatic behaviors, such as breathing and swallowing, or the startle response we feel when someone leaps out from behind a bush. Toward the center of the skull is a golf ball–

sized lump of tissue that is similar to what you might find inside the head of a fish, reptile, or mammal. This is the basal ganglia, an oval of cells that, for years, scientists didn't understand very well, except for suspicions that it played a role in diseases such as Parkinson's.

In the early 1990s, the MIT researchers began wondering if the basal ganglia might be integral to habits as well. They noticed that animals with injured basal ganglia suddenly developed problems with tasks such as learning how to run through mazes or remembering how to open food containers. They decided to experiment by employing new micro-technologies that allowed them to observe, in minute detail, what was occurring within the heads of rats as they performed dozens of routines. In surgery, each rat had what looked like a small joystick and dozens of tiny wires inserted into its skull. Afterward, the animal was placed into a T-shaped maze with chocolate at one end.

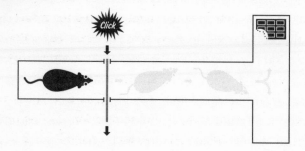

The maze was structured so that each rat was positioned behind a partition that opened when a loud click sounded. Initially, when a rat heard the click and saw the partition disappear, it would usually wander up and down the center aisle, sniffing in corners and scratching at walls. It appeared to smell the chocolate, but couldn't figure out how to find it. When it reached the top of the T, it often turned to the right, away from the chocolate, and then wandered left, sometimes pausing for no obvious reason. Eventually, most animals discovered the reward. But there was no discernible pattern in their meanderings. It seemed as if each rat was taking a leisurely, unthinking stroll.

The probes in the rats' heads, however, told a different story. While each animal wandered through the maze, its brain—and in particular, its basal ganglia—worked furiously. Each time a rat sniffed the air or scratched a wall, its brain exploded with activity, as if analyzing each new scent, sight, and sound. The rat was processing information the entire time it meandered.

The scientists repeated their experiment, again and again, watching how each rat's brain activity changed as it moved through the same route hundreds of times. A series of shifts slowly emerged. The rats stopped sniffing corners and making wrong turns. Instead, they zipped through the maze faster and faster. And within their brains, something unexpected occurred: As each rat learned how to navigate the maze, its mental activity *decreased*. As the route became more and more automatic, each rat started thinking less and less.

It was as if the first few times a rat explored the maze, its brain had to work at full power to make sense of all the new information. But after a few days of running the same route, the rat didn't need to scratch the walls or smell the air anymore, and so the brain activity associated with scratching and smelling ceased. It didn't need to choose which direction to turn, and so decision-making centers of the brain went quiet. All it had to do was recall the quickest path to the chocolate. Within a week, even the brain structures related to memory had quieted. The rat had internalized how to sprint through the maze to such a degree that it hardly needed to think at all.

But that internalization—run straight, hang a left, eat the chocolate—relied upon the basal ganglia, the brain probes indicated. This tiny, ancient neurological structure seemed to take over as the rat ran faster and faster and its brain worked less and less. The basal ganglia was central to recalling patterns and acting on them. The basal ganglia, in other words, stored habits even while the rest of the brain went to sleep.

To see this capacity in action, consider this graph, which shows activity within a rat's skull as it encounters the maze for the first time. Initially, the brain is working hard the entire time:

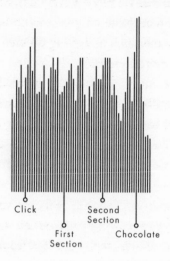

Click

First
Section

Second
Section

Chocolate

After a week, once the route is familiar and the scurrying has become a habit, the rat's brain settles down as it runs through the maze:

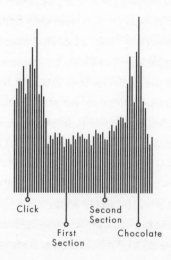

Click

First
Section

Second
Section

Chocolate

This process—in which the brain converts a sequence of actions into an automatic routine—is known as "chunking," and it's at the root of how habits form. There are dozens—if not hundreds—of behavioral chunks that we rely on every day. Some are simple: You automatically put toothpaste on your toothbrush before sticking it in your mouth. Some, such as getting dressed or making the kids' lunch, are a little more complex.

Others are so complicated that it's remarkable a small bit of tissue that evolved millions of years ago can turn them into habits at all. Take the act of backing your car out of the driveway. When you first learned to drive, the driveway required a major dose of concentration, and for good reason: It involves opening the garage, unlocking the car door, adjusting the seat, inserting the key in the ignition, turning it clockwise, moving the rearview and side mirrors and checking for obstacles, putting your foot on the brake, moving the gearshift into reverse, removing your foot from the brake, mentally estimating the distance between the garage and the street while keeping the wheels aligned and monitoring for oncoming traffic, calculating how reflected images in the mirrors translate into actual distances between the bumper, the garbage cans, and the hedges, all while applying slight pressure to the gas pedal and brake, and, most likely, telling your passenger to please stop fiddling with the radio.

Nowadays, however, you do all of that every time you pull onto the street with hardly any thought. The routine occurs by habit.

Millions of people perform this intricate ballet every morning, unthinkingly, because as soon as we pull out the car keys, our basal ganglia kicks in, identifying the habit we've stored in our brains related to backing an automobile into the street. Once that habit starts unfolding, our gray matter is free to quiet itself or chase other thoughts, which is why we have enough mental capacity to realize that Jimmy forgot his lunchbox inside.

Habits, scientists say, emerge because the brain is constantly looking for ways to save effort. Left to its own devices, the brain will try to make almost any routine into a habit, because habits allow our minds to ramp down more often. This effort-saving instinct is a huge advantage. An efficient brain requires less room, which makes for a smaller head, which makes childbirth easier and therefore causes fewer infant and mother deaths. An efficient brain also allows us to stop thinking constantly about basic behaviors, such as walking and choosing what to eat, so we can devote mental energy to inventing spears, irrigation systems, and, eventually, airplanes and video games.

But conserving mental effort is tricky, because if our brains power down at the wrong moment, we might fail to notice something important, such as a predator hiding in the bushes or a speeding car as we pull onto the street. So our basal ganglia have devised a clever system to determine when to let habits take over. It's something that happens whenever a chunk of behavior starts or ends.

To see how it works, look closely at the graph of the rat's neurological habit again. Notice that brain activity spikes at the beginning of the maze, when the rat hears the click before the partition starts moving, and again at the end, when it finds the chocolate.

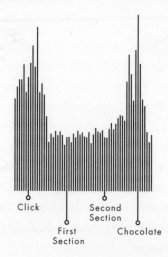

Those spikes are the brain's way of determining when to cede control to a habit, and which habit to use. From behind a partition, for instance, it's difficult for a rat to know if it's inside a familiar maze or an unfamiliar cupboard with a cat lurking outside. To deal with this uncertainty, the brain spends a lot of effort at the beginning of a habit looking for something—a cue—that offers a hint as to which pattern to use. From behind a partition, if a rat hears a click, it knows to use the maze habit. If it hears a meow, it chooses a different pattern. And at the end of the activity, when the reward appears, the brain shakes itself awake and makes sure everything unfolded as expected.

This process within our brains is a three-step loop. First, there is a *cue*, a trigger that tells your brain to go into automatic mode and which habit to use. Then there is the *routine*, which can be physical or mental or emotional. Finally, there is a *reward*, which helps your brain figure out if this particular loop is worth remembering for the future:

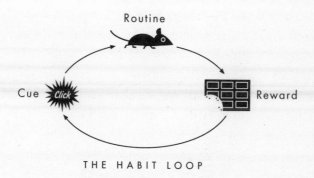

THE HABIT LOOP

Over time, this loop—cue, routine, reward; cue, routine, reward—becomes more and more automatic. The cue and reward become intertwined until a powerful sense of anticipation and craving emerges. Eventually, whether in a chilly MIT laboratory or your driveway, a habit is born.

● ● ●

Habits aren't destiny. As the next two chapters explain, habits can be ignored, changed, or replaced. But the reason the discovery of the habit loop is so important is that it reveals a basic truth: When a habit emerges, the brain stops fully participating in decision making. It stops working so hard, or diverts focus to other tasks. So unless you deliberately *fight* a habit—unless you find new routines—the pattern will unfold automatically.

However, simply understanding how habits work—learning the structure of the habit loop—makes them easier to control. Once you break a habit into its components, you can fiddle with the gears.

"We've done experiments where we trained rats to run down a maze until it was a habit, and then we extinguished the habit by changing the placement of the reward," Ann Graybiel, a scientist at MIT who oversaw many of the basal ganglia experiments, told me. "Then one day, we'll put the reward in the old place, and put in the rat, and, by golly, the old habit will reemerge right away. Habits never really disappear. They're encoded into the structures of our brain, and that's a huge advantage for us, because it would be awful if we had to relearn how to drive after every vacation. The problem is that your brain can't tell the difference between bad and good habits, and so if you have a bad one, it's always lurking there, waiting for the right cues and rewards."

This explains why it's so hard to create exercise habits, for instance, or change what we eat. Once we develop a routine of sitting on the couch, rather than running, or snacking whenever we pass a doughnut box, those patterns always remain inside our heads. By the same rule, though, if we learn to create new neurological routines that overpower those behaviors—if we take control of the habit loop—we can force those bad tendencies into the background, just as Lisa Allen did after her Cairo trip. And once someone creates a new pattern, studies have demonstrated, going for a jog or ignoring the doughnuts becomes as automatic as any other habit.

Without habit loops, our brains would shut down, overwhelmed by the minutiae of daily life. People whose basal ganglia are damaged by injury or disease often become mentally paralyzed. They have trouble performing basic activities, such as opening a door or deciding what to eat. They lose the ability to ignore insignificant details—one study, for example, found that patients with basal ganglia injuries couldn't recognize facial expressions, including fear and disgust, because they were perpetually uncertain about which part of the face to focus on. Without our basal ganglia, we lose access to the hundreds of habits we rely on every day. Did you pause this morning to decide whether to tie your left or right shoe first? Did you have trouble figuring out if you should brush your teeth before or after you showered?

Of course not. Those decisions are habitual, effortless. As long as your basal ganglia is intact and the cues remain constant, the behaviors will occur unthinkingly. (Though when you go on vacation, you may get dressed in different ways or brush your teeth at a different point in your morning routine without noticing it.)

At the same time, however, the brain's dependence on automatic routines can be dangerous. Habits are often as much a curse as a benefit.

Take Eugene, for instance. Habits gave him his life back after he lost his memory. Then they took everything away again.

III.

As Larry Squire, the memory specialist, spent more and more time with Eugene, he became convinced his patient was somehow learning new behaviors. Images of Eugene's brain showed that his basal ganglia had escaped injury from the viral encephalitis. Was it possible, the scientist wondered, that Eugene, even with severe brain damage, could still use the cue-routine-reward loop? Could this ancient neurological process explain how Eugene was able to walk around the block and find the jar of nuts in the kitchen?

To test if Eugene was forming new habits, Squire devised an experiment. He took sixteen different objects—bits of plastic and brightly colored pieces of toys—and glued them to cardboard rectangles. He then divided them into eight pairs: choice A and choice B. In each pairing, one piece of cardboard, chosen at random, had a sticker placed on the bottom that read "correct."

Eugene was seated at a table, given a pair of objects, and asked to choose one. Next, he was told to turn over his choice to see if there was a "correct" sticker underneath. This is a common way to measure memory. Since there are only sixteen objects, and they are always presented in the same eight pairings, most people can memorize which item is "correct" after a few rounds. Monkeys can memorize all the "correct" items after eight to ten days.

Eugene couldn't remember any of the "correct" items, no matter how many times he did the test. He repeated the experiment twice a week for months, looking at forty pairings each day.

"Do you know why you are here today?" a researcher asked at the beginning of one session a few weeks into the experiment.

"I don't think so," Eugene said.

"I'm going to show you some objects. Do you know why?"

"Am I supposed to describe them to you, or tell you what they are used for?" Eugene couldn't recollect the previous sessions at all.

But as the weeks passed, Eugene's performance improved. After twenty-eight days of training, Eugene was choosing the "correct" object 85 percent of the time. At thirty-six days, he was right 95 percent of the time. After one test, Eugene looked at the researcher, bewildered by his success.

"How am I doing this?" he asked her.

"Tell me what is going on in your head," the researcher said. "Do you say to yourself, 'I remember seeing that one'?"

"No," Eugene said. "It's here somehow or another"—he pointed to his head—"and the hand goes for it."

To Squire, however, it made perfect sense. Eugene was exposed to a cue: a pair of objects always presented in the same combination. There was a routine: He would choose one object and look to see if there was a sticker underneath, even if he had no idea why he felt compelled to turn the cardboard over. Then there was a reward: the satisfaction he received after finding a sticker proclaiming "correct." Eventually, a habit loop emerged.

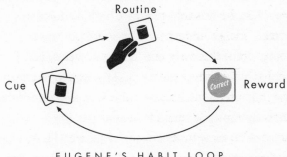

EUGENE'S HABIT LOOP

To make sure this pattern was, in fact, a habit, Squire conducted one more experiment. He took all sixteen items and put them in front of Eugene at the same time. He asked him to put all the "correct" objects into one pile.

Eugene had no idea where to begin. "Gosh sakes, how to remember this?" he asked. He reached for one object and started to turn it over. The experimenter stopped him. No, she explained. The task was to put the items in *piles*. Why was he trying to turn them over?

"That's just a habit, I think," he said.

He couldn't do it. The objects, when presented outside of the context of the habit loop, made no sense to him.

Here was the proof Squire was looking for. The experiments demonstrated that Eugene had the ability to form new habits, even when they involved tasks or objects he couldn't remember for more than a few seconds. This explained how Eugene managed to go for a walk every morning. The cues—certain trees on corners or the placement of particular mailboxes—were consistent every time he went outside, so though he couldn't recognize his house, his habits always guided him back to his front door. It also explained why Eugene would eat breakfast three or four times a day, even if he wasn't hungry. As long as the right cues were present—such as his radio or the morning light through his windows—he automatically followed the script dictated by his basal ganglia.

What's more, there were dozens of other habits in Eugene's life that no one noticed until they started looking for them. Eugene's daughter, for instance, would often stop by his house to say hello. She would talk to her father in the living room for a bit, then go into the kitchen to visit with her mother, and then leave, waving good-bye on her way out the door. Eugene, who had forgotten their earlier conversation by the time she left, would get angry—why was she leaving without chatting?—and then forget why he was upset. But the emotional habit had already started, and so his anger would persist, red hot and beyond his understanding, until it burned itself out.

"Sometimes he would bang the table or curse, and if you asked him why, he'd say 'I don't know, but I'm mad!'" Beverly told me. He would kick his chair, or snap at whoever came into the room. Then, a few minutes later, he would smile and talk about the weather. "It was like, once it started, he had to finish the frustration," she said.

Squire's new experiment also showed something else: that habits are surprisingly delicate. If Eugene's cues changed the slightest bit, his habits fell apart. The few times he walked around the block, for instance, and something was different—the city was doing street repairs or a windstorm had blown branches all over the sidewalk—Eugene would get lost, no matter how close he was to home, until a kind neighbor showed him the way to his door. If his daughter stopped to chat with him for ten seconds before she walked out, his anger habit never emerged.

Squire's experiments with Eugene revolutionized the scientific community's understanding of how the brain works by proving, once and for all, that it's possible to learn and make unconscious choices without remembering anything about the lesson or decision making. Eugene showed that habits, as much as memory and reason, are at the root of how we behave. We might not remember the experiences that create our habits, but once they are lodged within our brains they influence how we act—often without our realization.

● ● ●

Since Squire's first paper on Eugene's habits was published, the science of habit formation has exploded into a major field of study. Researchers at Duke, Harvard, UCLA, Yale, USC, Princeton, the University of Pennsylvania, and at schools in the United Kingdom, Germany, and the Netherlands, as well as corporate scientists working for Procter & Gamble, Microsoft, Google, and hundreds of other companies are focused on understanding the neurology and psychology of habits, their strengths and weaknesses, and why they emerge and how they can be changed.

Researchers have learned that cues can be almost anything, from a visual trigger such as a candy bar or a television commercial to a certain place, a time of day, an emotion, a sequence of thoughts, or the company of particular people. Routines can be incredibly complex or fantastically simple (some habits, such as those related to emotions, are measured in milliseconds). Rewards can range from food or drugs that cause physical sensations, to emotional payoffs, such as the feelings of pride that accompany praise or self-congratulation.

And in almost every experiment, researchers have seen echoes of Squire's discoveries with Eugene: Habits are powerful, but delicate. They can emerge outside our consciousness, or can be deliberately designed. They often occur without our permission, but can be re-shaped by fiddling with their parts. They shape our lives far more than we realize—they are so strong, in fact, that they cause our brains to cling to them at the exclusion of all else, including common sense.

In one set of experiments, for example, researchers affiliated with the National Institute on Alcohol Abuse and Alcoholism trained mice to press levers in response to certain cues until the behavior became a habit. The mice were always rewarded with food. Then, the scientists poisoned the food so that it made the animals violently ill, or electrified the floor, so that when the mice walked toward their reward they received a shock. The mice knew the food and cage were dangerous—when they were offered the poisoned pellets in a bowl or saw the electrified floor panels, they stayed away. When they saw their old cues, however, they unthinkingly pressed the lever and ate the food, or they walked across the floor, even as they vomited or jumped from the electricity. The habit was so ingrained the mice couldn't stop themselves.

It's not hard to find an analog in the human world. Consider fast food, for instance. It makes sense—when the kids are starving and you're driving home after a long day—to stop, just this once, at McDonald's or Burger King. The meals are inexpensive. It tastes so good. After all, one dose of processed meat, salty fries, and sugary soda poses a relatively small health risk, right? It's not like you do it all the time.

But habits emerge without our permission. Studies indicate that families usually don't *intend* to eat fast food on a regular basis. What happens is that a once a month pattern slowly becomes once a week, and then twice a week—as the cues and rewards create a habit—until the kids are consuming an unhealthy amount of hamburgers and fries. When researchers at the University of North Texas and Yale tried to understand why families gradually increased their fast food consumption, they found a series of cues and rewards that most customers never knew were influencing their behaviors. They discovered the habit loop.

Every McDonald's, for instance, looks the same—the company deliberately tries to standardize stores' architecture and what employees say to customers, so everything is a consistent cue to trigger eating routines. The foods at some chains are specifically engineered to deliver immediate rewards—the fries, for instance, are designed to begin disintegrating the moment they hit your tongue, in order to deliver a hit of salt and grease as fast as possible, causing your pleasure centers to light up and your brain to lock in the pattern. All the better for tightening the habit loop.

However, even these habits are delicate. When a fast food restaurant closes down, the families that previously ate there will often start having dinner at home, rather than seek out an alternative location. Even small shifts can end the pattern. But since we often don't recognize these habit loops as they grow, we are blind to our ability to control them. By learning to observe the cues and rewards, though, we can change the routines.

IV.

By 2000, seven years after Eugene's illness, his life had achieved a kind of equilibrium. He went for a walk every morning. He ate what he wanted, sometimes five or six times a day. His wife knew that as long as the television was tuned to the History Channel, Eugene would settle into his plush chair and watch it regardless of whether it was airing reruns or new programs. He couldn't tell the difference.

As he got older, however, Eugene's habits started impacting his life in negative ways. He was sedentary, sometimes watching television for hours at a time because he never grew bored with the shows. His physicians became worried about his heart. The doctors told Beverly to keep him on a strict diet of healthy foods. She tried, but it was difficult to influence how frequently he ate or what he consumed. He never recalled her admonitions. Even if the refrigerator was stocked with fruits and vegetables, Eugene would root around until he found the bacon and eggs. That was his routine. And as Eugene aged and his bones became more brittle, the doctors said he needed to be more careful walking around. In his mind, however, Eugene was twenty years younger. He never remembered to step carefully.

"All my life I was fascinated by memory," Squire told me. "Then I met E.P., and saw how rich life can be even if you can't remember it. The brain has this amazing ability to find happiness even when the memories of it are gone.

"It's hard to turn that off, though, which ultimately worked against him."

Beverly tried to use her understanding of habits to help Eugene avoid problems as he aged. She discovered that she could short-circuit some of his worst patterns by inserting new cues. If she didn't keep bacon in the fridge, Eugene wouldn't eat multiple, unhealthy breakfasts. When she put a salad next to his chair, he would sometimes pick at it, and as the meal became a habit, he stopped searching the kitchen for treats. His diet gradually improved.

Despite these efforts, however, Eugene's health still declined. One spring day, Eugene was watching television when he suddenly shouted. Beverly ran in and saw him clutching his chest. She called an ambulance. At the hospital, they diagnosed a minor heart attack. By then the pain had passed and Eugene was fighting to get off his gurney. That night, he kept pulling off the monitors attached to his chest so he could roll over and sleep. Alarms would blare and nurses would rush in. They tried to get him to quit fiddling with the sensors by taping the leads in place and telling him they would use restraints if he continued fussing. Nothing worked. He forgot the threats as soon as they were issued.

Then his daughter told a nurse to try complimenting him on his willingness to sit still, and to repeat the compliment, over and over, each time she saw him. "We wanted to, you know, get his pride involved," his daughter, Carol Rayes, told me. "We'd say, 'Oh, Dad, you're really doing something important for science by keeping these doodads in place.'" The nurses started to dote on him. He loved it. After a couple of days, he did whatever they asked. Eugene returned home a week later.

Then, in the fall of 2008, while walking through his living room, Eugene tripped on a ledge near the fireplace, fell, and broke his hip. At the hospital, Squire and his team worried that he would have panic attacks because he wouldn't know where he was. So they left notes by his bedside explaining what had happened and posted photos of his children on the walls. His wife and kids came every day.

Eugene, however, never grew worried. He never asked why he was in the hospital. "He seemed at peace with all the uncertainty by that point," said Squire. "It had been fifteen years since he had lost his memory. It was as if part of his brain knew there were some things he would never understand and was okay with that."

Beverly came to the hospital every day. "I spent a long time talking to him," she said. "I told him that I loved him, and about our kids and what a good life we had. I pointed to the pictures and talked about how much he was adored. We were married for fifty-seven years, and forty-two of those were a real, normal marriage. Sometimes it was hard, because I wanted my old husband back so much. But at least I knew he was happy."

A few weeks later, his daughter came to visit. "What's the plan?" Eugene asked when she arrived. She took him outside in a wheelchair, onto the hospital's lawn. "It's a beautiful day," Eugene said. "Pretty nice weather, huh?" She told him about her kids and they played with a dog. She thought he might be able to come home soon. The sun was going down. She started to get ready to take him inside.

Eugene looked at her.

"I'm lucky to have a daughter like you," he said. She was caught off-guard. She couldn't remember the last time he had said something so sweet.

"I'm lucky that you're my dad," she told him.

"Gosh, it's a beautiful day," he said. "What do you think about the weather?"

That night, at one o'clock in the morning, Beverly's phone rang. The doctor said Eugene had suffered a massive heart attack and the staff had done everything possible, but hadn't been able to revive him. He was gone. After his death, he would be celebrated by researchers, the images of his brain studied in hundreds of labs and medical schools.

"I know he would have been really proud to know how much he contributed to science," Beverly told me. "He told me once, pretty soon after we got married, that he wanted to do something important with his life, something that mattered. And he did. He just never remembered any of it."

2

THE CRAVING BRAIN

How to Create New Habits

I.

One day in the early 1900s, a prominent American executive named Claude C. Hopkins was approached by an old friend with a new business idea. The friend had discovered an amazing product, he explained, that he was convinced would be a hit. It was a toothpaste, a minty, frothy concoction he called "Pepsodent." There were some dicey investors involved—one of them had a string of busted land deals; another, it was rumored, was connected to the mob—but this venture, the friend promised, was going to be huge. If, that is, Hopkins would consent to help design a national promotional campaign.

Hopkins, at the time, was at the top of a booming industry that had hardly existed a few decades earlier: advertising. Hopkins was the man who had convinced Americans to buy Schlitz beer by boasting that the company cleaned their bottles "with live steam," while neglecting to mention that every other company used the exact same method. He had seduced millions of women into purchasing Palmolive soap by proclaiming that Cleopatra had washed with it, despite the sputtering protests of outraged historians. He had made Puffed Wheat famous by saying that it was "shot from guns" until the grains puffed "to eight times normal size." He had turned dozens of previously unknown products—Quaker Oats, Goodyear tires, the Bissell carpet sweeper, Van Camp's pork and beans—into household names. And in the process, he had made himself so rich that his best-selling autobiography, *My Life in Advertising*, devoted long passages to the difficulties of spending so much money.

Claude Hopkins was best known for a series of rules he coined explaining how to create new habits among consumers. These rules would transform industries and eventually became conventional wisdom among marketers, educational reformers, public health professionals, politicians, and CEOs. Even today, Hopkins's rules influence everything from how we buy cleaning supplies to the tools governments use for eradicating disease. They are fundamental to creating any new routine.

However, when his old friend approached Hopkins about Pepsodent, the ad man expressed only mild interest. It was no secret that the health of Americans' teeth was in steep decline. As the nation had become wealthier, people had started buying larger amounts of sugary, processed foods. When the government started drafting men for World War I, so many recruits had rotting teeth that officials said poor dental hygiene was a national security risk.

Yet as Hopkins knew, selling toothpaste was financial suicide. There was already an army of door-to-door salesmen hawking dubious tooth powders and elixirs, most of them going broke.

The problem was that hardly anyone bought toothpaste because, despite the nation's dental problems, hardly anyone brushed their teeth.

So Hopkins gave his friend's proposal a bit of thought, and then declined. He'd stick with soaps and cereals, he said. "I did not see a way to educate the laity in technical tooth-paste theories," Hopkins explained in his autobiography. The friend, however, was persistent. He came back again and again, appealing to Hopkins's considerable ego until, eventually, the ad man gave in.

"I finally agreed to undertake the campaign if he gave me a six months' option on a block of stock," Hopkins wrote. The friend agreed.

It would be the wisest financial decision of Hopkins's life.

Within five years of that partnership, Hopkins turned Pepsodent into one of the best-known products on earth and, in the process, helped create a toothbrushing habit that moved across America with startling speed. Soon, everyone from Shirley Temple to Clark Gable was bragging about their "Pepsodent smile." By 1930, Pepsodent was sold in China, South Africa, Brazil, Germany, and almost anywhere else Hopkins could buy ads. A decade after the first Pepsodent campaign, pollsters found that toothbrushing had become a ritual for more than half the American population. Hopkins had helped establish toothbrushing as a daily activity.

The secret to his success, Hopkins would later boast, was that he had found a certain kind of cue and reward that fueled a particular habit. It's an alchemy so powerful that even today the basic principles are still used video game designers, food companies, hospitals, and millions of salesmen around the world. Eugene Pauly taught us about the habit loop, but it was Claude Hopkins that showed how new habits can be cultivated and grown.

So what, exactly, did Hopkins do?

He created a craving. And that craving, it turns out, is what makes cues and rewards work. That craving is what powers the habit loop.

● ● ●

Throughout his career, one of Claude Hopkins's signature tactics was to find simple triggers to convince consumers to use his products every day. He sold Quaker Oats, for instance, as a breakfast cereal that could provide energy for twenty-four hours—but only if you ate a bowl every morning. He hawked tonics that cured stomachaches, joint pain, bad skin, and "womanly problems"—but only if you drank the medicine at symptoms' first appearance. Soon, people were devouring oatmeal at daybreak and chugging from little brown bottles whenever they felt a hint of fatigue, which, as luck would have it, often happened at least once a day.

To sell Pepsodent, then, Hopkins needed a trigger that would justify the toothpaste's daily use. He sat down with a pile of dental textbooks. "It was dry reading," he later wrote. "But in the middle of one book I found a reference to the mucin plaques on teeth, which I afterward called 'the film.' That gave me an appealing idea. I resolved to advertise this toothpaste as a creator of beauty. To deal with that cloudy film."

In focusing on tooth film, Hopkins was ignoring the fact that this same film has always covered people's teeth and hadn't seemed to bother anyone. The film is a naturally occurring membrane that builds up on teeth regardless of what you eat or how often you brush. People had never paid much attention to it, and there was little reason why they should: You can get rid of the film by eating an apple, running your finger over your teeth, brushing, or vigorously swirling liquid around your mouth. Toothpaste didn't do anything to help remove the film. In fact, one of the leading dental researchers of the time said that all toothpastes—particularly Pepsodent—were worthless.

That didn't stop Hopkins from exploiting his discovery. Here, he decided, was a cue that could trigger a habit. Soon, cities were plastered with Pepsodent ads.

"Just run your tongue across your teeth," read one. "*You'll feel a film*—that's what makes your teeth look 'off color' and invites decay."

"Note how many pretty teeth are seen everywhere," read another ad, featuring smiling beauties. "Millions are using a new method of teeth cleansing. Why would any woman have dingy film on her teeth? Pepsodent removes the film!"

The brilliance of these appeals was that they relied upon a cue—tooth film—that was universal and impossible to ignore. Telling someone to run their tongue across their teeth, it turned out, was likely to cause them to run their tongue across their teeth. And when they did, they were likely to feel a film. Hopkins had found a cue that was simple, had existed for ages, and was so easy to trigger that an advertisement could cause people to comply automatically.

Moreover, the reward, as Hopkins envisioned it, was even more enticing. Who, after all, doesn't want to be more beautiful? Who doesn't want a prettier smile? Particularly when all it takes is a quick brush with Pepsodent?

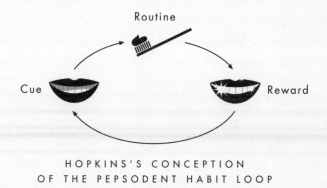

HOPKINS'S CONCEPTION
OF THE PEPSODENT HABIT LOOP

After the campaign launched, a quiet week passed. Then two. In the third week, demand exploded. There were so many orders for Pepsodent that the company couldn't keep up. In three years, the product went international, and Hopkins was crafting ads in Spanish, German, and Chinese. Within a decade, Pepsodent was one of the top-selling goods in the world, and remained America's best-selling toothpaste for more than thirty years.

Before Pepsodent appeared, only 7 percent of Americans had a tube of toothpaste in their medicine chests. A decade after Hopkins's ad campaign went nationwide, that number had jumped to 65 percent. By the end of World War II, the military downgraded concerns about recruits' teeth because so many soldiers were brushing every day.

"I made for myself a million dollars on Pepsodent," Hopkins wrote a few years after the product appeared on shelves. The key, he said, was that he had "learned the right human psychology." That psychology was grounded in two basic rules:

First, find a simple and obvious cue.

Second, clearly define the rewards.

If you get those elements right, Hopkins promised, it was like magic. Look at Pepsodent: He had identified a cue—tooth film—and a reward—beautiful teeth—that had persuaded millions to start a daily ritual. Even today, Hopkins's rules are a staple of marketing textbooks and the foundation of millions of ad campaigns.

And those same principles have been used to create thousands of other habits—often without people realizing how closely they are hewing to Hopkins's formula. Studies of people who have successfully started new exercise routines, for instance, show they are more likely to stick with a workout plan if they choose a specific cue, such as running as soon as they get home from work, and a clear reward, such as a beer or an evening of guilt-free television. Research on dieting says creating new food habits requires a predetermined cue—such as planning menus in advance—and simple rewards for dieters when they stick to their intentions.

"The time has come when advertising has in some hands reached the status of a science," Hopkins wrote. "Advertising, once a gamble, has thus become, under able direction, one of the safest of business ventures."

It's quite a boast. However, it turns out that Hopkins's two rules aren't enough. There's also a third rule that must be satisfied to create a habit—a rule so subtle that Hopkins himself relied on it without knowing it existed. It explains everything from why it's so hard to ignore a box of doughnuts to how a morning jog can become a nearly effortless routine.

II.

The scientists and marketing executives at Procter & Gamble were gathered around a beat-up table in a small, windowless room, reading the transcript of an interview with a woman who owned nine cats, when one of them finally said what everyone was thinking.

"If we get fired, what exactly happens?" she asked. "Do security guards show up and walk us out, or do we get some kind of warning beforehand?"

The team's leader, a onetime rising star within the company named Drake Stimson, stared at her.

"I don't know," he said. His hair was a mess. His eyes were tired. "I never thought things would get this bad. They told me running this project was a promotion."

It was 1996, and the group at the table was finding out, despite Claude Hopkins's assertions, how utterly unscientific the process of selling something could become. They all worked for one of the largest consumer goods firms on earth, the company behind Pringles potato chips, Oil of Olay, Bounty paper towels, CoverGirl cosmetics, Dawn, Downy, and Duracell, as well as dozens of other brands. P&G collected more data than almost any other merchant on earth and relied on complex statistical methods to craft their marketing campaigns. The firm was incredibly good at figuring out how to sell things. In the clothes-washing market alone, P&G's products cleaned one out of every two laundry loads in America. Its revenues topped $35 billion per year.

However, Stimson's team, which had been entrusted with designing the ad campaign for one of P&G's most promising new products, was on the brink of failure. The company had spent millions of dollars developing a spray that could remove bad smells from almost any fabric. And the researchers in that tiny, windowless room had no idea how to get people to buy it.

The spray had been created about three years earlier, when one of P&G's chemists was working with a substance called hydroxypropyl beta cyclodextrin, or HPBCD, in a laboratory. The chemist was a smoker. His clothes usually smelled like an ashtray. One day, after working with HPBCD, his wife greeted him at the door when he got home.

"Did you quit smoking?" she asked him.

"No," he said. He was suspicious. She had been harassing him to give up cigarettes for years. This seemed like some kind of reverse psychology trickery.

"You don't smell like smoke, is all," she said.

The next day, he went back to the lab and started experimenting with HPBCD and various scents. Soon, he had hundreds of vials containing fabrics that smelled like wet dogs, cigars, sweaty socks, Chinese food, musty shirts, and dirty towels. When he put HPBCD in water and sprayed it on the samples, the scents were drawn into the chemical's molecules. After the mist dried, the smell was gone.

When the chemist explained his findings to P&G's executives, they were ecstatic. For years, market research had said that consumers were clamoring for something that could get rid of bad smells— not mask them, but eradicate them altogether. When one team of researchers had interviewed customers, they found that many of them left their blouses or slacks outside after a night at a bar or party. "My clothes smell like cigarettes when I get home, but I don't want to pay for dry cleaning every time I go out," one woman said.

P&G, sensing an opportunity, launched a top-secret project to turn HPBCD into a viable product. They spent millions perfecting the formula, finally producing a colorless, odorless liquid that could wipe out almost any foul odor. The science behind the spray was so advanced that NASA would eventually use it to clean the interiors of shuttles after they returned from space. The best part was that it was cheap to manufacture, didn't leave stains, and could make any stinky couch, old jacket, or stained car interior smell, well, scentless. The project had been a major gamble, but P&G was now poised to earn billions—if they could come up with the right marketing campaign.

They decided to call it Febreze, and asked Stimson, a thirty-one-year-old wunderkind with a background in math and psychology, to lead the marketing team. Stimson was tall and handsome, with a strong chin, a gentle voice, and a taste for high-end meals. ("I'd rather my kids smoked weed than ate in McDonald's," he once told a colleague.) Before joining P&G, he had spent five years on Wall Street building mathematical models for choosing stocks. When he relocated to Cincinnati, where P&G was headquartered, he was tapped to help run important business lines, including Bounce fabric softener and Downy dryer sheets. But Febreze was different. It was a chance to launch an entirely new category of product—to add something to a consumer's shopping cart that had never been there before. All Stimson needed to do was figure out how to make Febreze into a habit, and the product would fly off the shelves. How tough could that be?

Stimson and his colleagues decided to introduce Febreze in a few test markets—Phoenix, Salt Lake City, and Boise. They flew in and handed out samples, and then asked people if they could come by their homes. Over the course of two months, they visited hundreds of households. Their first big breakthrough came when they visited a park ranger in Phoenix. She was in her late twenties and lived by herself. Her job was to trap animals that wandered out of the desert. She caught coyotes, raccoons, the occasional mountain lion. And skunks. Lots and lots of skunks. Which often sprayed her when they were caught.

"I'm single, and I'd like to find someone to have kids with," the ranger told Stimson and his colleagues while they sat in her living room. "I go on a lot of dates. I mean, I think I'm attractive, you know? I'm smart and I feel like I'm a good catch."

But her love life was crippled, she explained, because everything in her life smelled like skunk. Her house, her truck, her clothing, her boots, her hands, her curtains. Even her bed. She had tried all sorts of cures. She bought special soaps and shampoos. She burned candles and used expensive carpet shampooing machines. None of it worked.

"When I'm on a date, I'll get a whiff of something that smells like skunk and I'll start obsessing about it," she told them. "I'll start wondering, does he smell it? What if I bring him home and he wants to leave?

"I went on four dates last year with a really nice guy, a guy I really liked, and I waited forever to invite him to my place. Eventually, he came over, and I thought everything was going really well. Then the next day, he said he wanted to 'take a break.' He was really polite about it, but I keep wondering, was it the smell?"

"Well, I'm glad you got a chance to try Febreze," Stimson said. "How'd you like it?"

She looked at him. She was crying.

"I want to thank you," she said. "This spray has changed my life."

After she had received samples of Febreze, she had gone home and sprayed her couch. She sprayed the curtains, the rug, the bedspread, her jeans, her uniform, the interior of her car. The bottle ran out, so she got another one, and sprayed everything else.

"I've asked all of my friends to come over," the woman said. "They can't smell it anymore. The skunk is gone."

By now, she was crying so hard that one of Stimson's colleagues was patting her on the shoulder. "Thank you so much," the woman said. "I feel so free. Thank you. This product is so important."

Stimson sniffed the air inside her living room. He couldn't smell anything. *We're going to make a fortune with this stuff,* he thought.

● ● ●

Stimson and his team went back to P&G headquarters and started reviewing the marketing campaign they were about to roll out. The key to selling Febreze, they decided, was conveying that sense of relief the park ranger felt. They had to position Febreze as something that would allow people to rid themselves of embarrassing smells. All of them were familiar with Claude Hopkins's rules, or the modern incarnations that filled business school textbooks. They wanted to keep the ads simple: Find an obvious cue and clearly define the reward.

They designed two television commercials. The first showed a woman talking about the smoking section of a restaurant. Whenever she eats there, her jacket smells like smoke. A friend tells her if she uses Febreze, it will eliminate the odor. The cue: the smell of cigarettes. The reward: odor eliminated from clothes. The second ad featured a woman worrying about her dog, Sophie, who always sits on the couch. "Sophie will always smell like Sophie," she says, but with Febreze, "now my furniture doesn't have to." The cue: pet smells, which are familiar to the seventy million households with animals. The reward: a house that doesn't smell like a kennel.

Stimson and his colleagues began airing the advertisements in 1996 in the same test cities. They gave away samples, put advertisements in mailboxes, and paid grocers to build mountains of Febreze near cash registers. Then they sat back, anticipating how they would spend their bonuses.

A week passed. Then two. A month. Two months. Sales started small—and got smaller. Panicked, the company sent researchers into stores to see what was happening. Shelves were filled with Febreze bottles that had never been touched. They started visiting housewives who had received free samples.

"Oh, yes!" one of them told a P&G researcher. "The spray! I remember it. Let's see." The woman got down on her knees in the kitchen and started rooting through the cabinet underneath the sink. "I used it for a while, but then I forgot about it. I think it's back here somewhere." She stood up. "Maybe it's in the closet?" She walked over and pushed aside some brooms. "Yes! Here it is! In the back! See? It's still almost full. Did you want it back?"

Febreze was a dud.

For Stimson, this was a disaster. Rival executives in other divisions sensed an opportunity in his failure. He heard whispers that some people were lobbying to kill Febreze and get him reassigned to Nicky Clarke hair products, the consumer goods equivalent of Siberia.

One of P&G's divisional presidents called an emergency meeting and announced they had to cut their losses on Febreze before board members started asking questions. Stimson's boss stood up and made an impassioned plea. "There's still a chance to turn everything around," he said. "At the very least, let's ask the PhDs to figure out what's going on." P&G had recently snapped up scientists from Stanford, Carnegie Mellon, and elsewhere who were supposed experts in consumer psychology. The division's president agreed to give the product a little more time.

So a new group of researchers joined Stimson's team and started conducting more interviews. Their first inkling of why Febreze was failing came when they visited a woman's home outside Phoenix. They could smell her nine cats before they went inside. The house's interior, however, was clean and organized. She was somewhat of a neat freak, the woman explained. She vacuumed every day and didn't like to open her windows, since the wind blew in dust. When Stimson and the scientists walked into her living room, where the cats lived, the scent was so overpowering that one of them gagged.

"What do you do about the cat smell?" a scientist asked the woman.

"It's usually not a problem," she said.

"How often do you notice a smell?"

"Oh, about once a month," the woman replied.

The researchers looked at one another.

"Do you smell it now?" a scientist asked.

"No," she said.

The same pattern played out in dozens of other smelly homes the researchers visited. People couldn't detect most of the bad smells in their lives. If you live with nine cats, you become desensitized to their scent. If you smoke cigarettes, it damages your olfactory capacities so much that you can't smell smoke anymore. Scents are strange; even the strongest fade with constant exposure. That's why no one was using Febreze, Stimson realized. The product's cue—the thing that was supposed to trigger daily use—was hidden from the people who needed it most. Bad scents simply weren't noticed frequently enough to trigger a regular habit. As a result, Febreze ended up in the back of a closet. The people with the greatest proclivity to use the spray never smelled the odors that should have reminded them the living room needed a spritz.

Stimson's team went back to headquarters and gathered in the windowless conference room, rereading the transcript of the woman with nine cats. The psychologist asked what happens if you get fired. Stimson put his head in his hands. If he couldn't sell Febreze to a woman with nine cats, he wondered, who *could* he sell it to? How do you build a new habit when there's no cue to trigger usage, and when the consumers who most need it don't appreciate the reward?

III.

The laboratory belonging to Wolfram Schultz, a professor of neuroscience at the University of Cambridge, is not a pretty place. His desk has been alternately described by colleagues as a black hole where documents are lost forever and a petri dish where organisms can grow, undisturbed and in wild proliferation, for years. When Schultz needs to clean something, which is uncommon, he doesn't use sprays or cleansers. He wets a paper towel and wipes hard. If his clothes smell like smoke or cat hair, he doesn't notice. Or care.

However, the experiments that Schultz has conducted over the past twenty years have revolutionized our understanding of how cues, rewards, and habits interact. He has explained why some cues and rewards have more power than others, and has provided a scientific road map that explains why Pepsodent was a hit, how some dieters and exercise buffs manage to change their habits so quickly, and—in the end—what it took to make Febreze sell.

In the 1980s, Schultz was part of a group of scientists studying the brains of monkeys as they learned to perform certain tasks, such as pulling on levers or opening clasps. Their goal was to figure out which parts of the brain were responsible for new actions.

"One day, I noticed this thing that is interesting to me," Schultz told me. He was born in Germany and now, when he speaks English, sounds a bit like Arnold Schwarzenegger if the Terminator were a member of the Royal Society. "A few of the monkeys we watched loved apple juice, and the other monkeys loved grape juice, and so I began to wonder, what is going on inside those little monkey heads? Why do different rewards affect the brain in different ways?"

Schultz began a series of experiments to decipher how rewards work on a neurochemical level. As technology progressed, he gained access, in the 1990s, to devices similar to those used by the researchers at MIT. Rather than rats, however, Schultz was interested in monkeys like Julio, an eight-pound macaque with hazel eyes who had a very thin electrode inserted into his brain that allowed Schultz to observe neuronal activity as it occurred.

One day, Schultz positioned Julio on a chair in a dimly lit room and turned on a computer monitor. Julio's job was to touch a lever whenever colored shapes—small yellow spirals, red squiggles, blue lines—appeared on the screen. If Julio touched the lever when a shape appeared, a drop of blackberry juice would run down a tube hanging from the ceiling and onto the monkey's lips.

Julio liked blackberry juice.

At first, Julio was only mildly interested in what was happening on the screen. He spent most of his time trying to squirm out of the chair. But once the first dose of juice arrived, Julio became very focused on the monitor. As the monkey came to understand, through dozens of repetitions, that the shapes on the screen were a cue for a routine (touch the lever) that resulted in a reward (blackberry juice), he started staring at the screen with a laserlike intensity. He didn't squirm. When a yellow squiggle appeared, he went for the lever. When a blue line flashed, he pounced. And when the juice arrived, Julio would lick his lips contentedly.

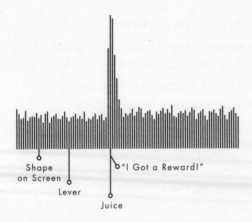

Shape on Screen

Lever

Juice

"I Got a Reward!"

JULIO'S REWARD RESPONSE WHEN
HE RECEIVES THE JUICE

As Schultz monitored the activity within Julio's brain, he saw a pattern emerge. Whenever Julio received his reward, his brain activity would spike in a manner that suggested he was experiencing happiness. A transcript of that neurological activity shows what it looks like when a monkey's brain says, in essence, "I got a reward!"

Schultz took Julio through the same experiment again and again, recording the neurological response each time. Whenever Julio received his juice, the "I got a reward!" pattern appeared on the computer attached to the probe in the monkey's head. Gradually, from a neurological perspective, Julio's behavior became a habit.

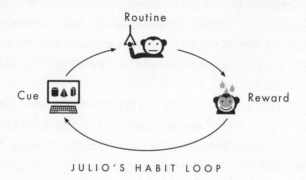

JULIO'S HABIT LOOP

What was most interesting to Schultz, however, was how things changed as the experiment proceeded. As the monkey became more and more practiced at the behavior—as the habit became stronger and stronger—Julio's brain began *anticipating* the blackberry juice. Schultz's probes started recording the "I got a reward!" pattern the instant Julio saw the shapes on the screen, *before* the juice arrived:

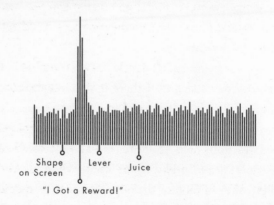

NOW, JULIO'S REWARD RESPONSE
OCCURS BEFORE THE JUICE ARRIVES

In other words, the shapes on the monitor had become a cue not just for pulling a lever, but also for a pleasure response inside the monkey's brain. Julio started expecting his reward as soon as he saw the yellow spirals and red squiggles.

Then Schultz adjusted the experiment. Previously, Julio had received juice as soon as he touched the lever. Now, sometimes, the juice didn't arrive at all, even if Julio performed correctly. Or it would arrive after a slight delay. Or it would be watered down until it was only half as sweet.

When the juice didn't arrive or was late or diluted, Julio would get angry and make unhappy noises, or become mopey. And within Julio's brain, Schultz watched a new pattern emerge: craving. When Julio anticipated juice but didn't receive it, a neurological pattern associated with desire and frustration erupted inside his skull. When Julio saw the cue, he started anticipating a juice-fueled joy. But if the juice didn't arrive, that joy became a craving that, if unsatisfied, drove Julio to anger or depression.

Researchers in other labs have found similar patterns. Other monkeys were trained to anticipate juice whenever they saw a shape on a screen. Then, researchers tried to distract them. They opened the lab's door, so the monkeys could go outside and play with their friends. They put food in a corner, so the monkeys could eat if they abandoned the experiment.

For those monkeys who hadn't developed a strong habit, the distractions worked. They slid out of their chairs, left the room, and never looked back. They hadn't learned to crave the juice. However, once a monkey had developed a habit—once its brain *anticipated* the reward—the distractions held no allure. The animal would sit there, watching the monitor and pressing the lever, over and over again, regardless of the offer of food or the opportunity to go outside. The anticipation and sense of craving was so overwhelming that the monkeys stayed glued to their screens, the same way a gambler will play slots long after he's lost his winnings.

This explains why habits are so powerful: They create neurological cravings. Most of the time, these cravings emerge so gradually that we're not really aware they exist, so we're often blind to their influence. But as we associate cues with certain rewards, a subconscious craving emerges in our brains that starts the habit loop spinning. One researcher at Cornell, for instance, found how powerfully food and scent cravings can affect behavior when he noticed how Cinnabon stores were positioned inside shopping malls. Most food sellers locate their kiosks in food courts, but Cinnabon tries to locate their stores *away* from other food stalls. Why? Because Cinnabon executives want the smell of cinnamon rolls to waft down hallways and around corners uninterrupted, so that shoppers will start subconsciously craving a roll. By the time a consumer turns a corner and sees the Cinnabon store, that craving is a roaring monster inside his head and he'll reach, unthinkingly, for his wallet. The habit loop is spinning because a sense of craving has emerged.

"There is nothing programmed into our brains that makes us see a box of doughnuts and automatically want a sugary treat," Schultz told me. "But once our brain learns that a doughnut box contains yummy sugar and other carbohydrates, it will start *anticipating* the sugar high. Our brains will push us toward the box. Then, if we don't eat the doughnut, we'll feel disappointed."

To understand this process, consider how Julio's habit emerged. First, he saw a shape on the screen:

Cue

Over time, Julio learned that the appearance of the shape meant it was time to execute a routine. So he touched the lever:

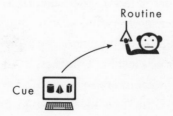

As a result, Julio received a drop of blackberry juice.

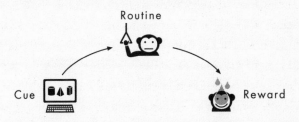

That's basic learning. The habit only emerges once Julio begins *craving* the juice when he sees the cue. Once that craving exists, Julio will act automatically. He'll follow the habit:

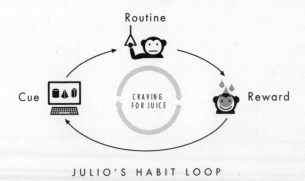

JULIO'S HABIT LOOP

This is how new habits are created: by putting together a cue, a routine, and a reward, and then cultivating a craving that drives the loop. Take, for instance, smoking. When a smoker sees a cue—say, a pack of Marlboros—her brain starts anticipating a hit of nicotine.

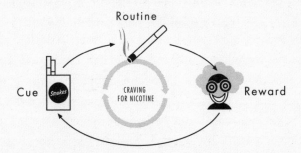

Just the sight of cigarettes is enough for the brain to crave a nicotine rush. If it doesn't arrive, the craving grows until the smoker reaches, unthinkingly, for a Marlboro.

Or take email. When a computer chimes or a smartphone vibrates with a new message, the brain starts anticipating the momentary distraction that opening an email provides. That expectation, if unsatisfied, can build until a meeting is filled with antsy executives checking their buzzing BlackBerrys under the table, even if they know it's probably only their latest fantasy football results. (On the other hand, if someone disables the buzzing—and, thus, removes the cue—people can work for hours without thinking to check their in-boxes.)

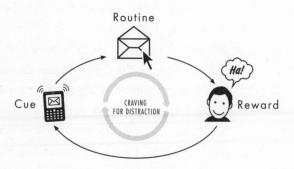

Scientists have studied the brains of alcoholics, smokers, and overeaters and have measured how their neurology—the structures of their brains and the flow of neurochemicals inside their skulls—changes as their cravings became ingrained. Particularly strong habits, wrote two researchers at the University of Michigan, produce addiction-like reactions so that "wanting evolves into obsessive craving" that can force our brains into autopilot, "even in the face of strong disincentives, including loss of reputation, job, home, and family."

However, these cravings don't have complete authority over us. As the next chapter explains, there are mechanisms that can help us ignore the temptations. But to overpower the habit, we must recognize which craving is driving the behavior. If we're not conscious of the anticipation, then we're like the shoppers who wander, as if drawn by an unseen force, into Cinnabon.

● ● ●

To understand the power of cravings in creating habits, consider how exercise habits emerge. In 2002 researchers at New Mexico State University wanted to understand why people habitually exercise. They studied 266 individuals, most of whom worked out at least three times a week. What they found was that many of them had started running or lifting weights almost on a whim, or because they suddenly had free time or wanted to deal with unexpected stresses in their lives. However, the reason they *continued*—why it became a habit—was because of a specific reward they started to crave.

In one group, 92 percent of people said they habitually exercised because it made them "feel good"—they grew to expect and crave the endorphins and other neurochemicals a workout provided. In another group, 67 percent of people said that working out gave them a sense of "accomplishment"—they had come to crave a regular sense of triumph from tracking their performances, and that self-reward was enough to make the physical activity into a habit.

If you want to start running each morning, it's essential that you choose a simple cue (like always lacing up your sneakers before breakfast or leaving your running clothes next to your bed) and a clear reward (such as a midday treat, a sense of accomplishment from recording your miles, or the endorphin rush you get from a jog). But countless studies have shown that a cue and a reward, on their own, aren't enough for a new habit to last. Only when your brain starts *expecting* the reward—craving the endorphins or sense of accomplishment—will it become automatic to lace up your jogging shoes each morning. The cue, in addition to triggering a routine, must also trigger a craving for the reward to come.

"Let me ask you about a problem I have," I said to Wolfram Schultz,

the neuroscientist, after he explained to me how craving emerges. "I have a two-year-old, and when I'm home feeding him dinner—chicken nuggets and stuff like that—I'll reach over and eat one myself without thinking about it. It's a habit. And now I'm gaining weight."

"Everybody does that," Schultz said. He has three children of his own, all adults now. When they were young, he would pick at their dinners unthinkingly. "In some ways," he told me, "we're like the monkeys. When we see the chicken or fries on the table, our brains begin anticipating that food, even if we're not hungry. Our brains are craving them. Frankly, I don't even *like* this kind of food, but suddenly, it's hard to fight this urge. And as soon as I eat it, I feel this rush of pleasure as the craving is satisfied. It's humiliating, but that's how habits work."

"I guess I should be thankful," he said, "because the same process has let me create good habits. I work hard because I expect pride from a discovery. I exercise because I expect feeling good afterward. I just wish I could pick and choose better."

IV.

After their disastrous interview with the cat woman, Drake Stimson's team at P&G started looking outside the usual channels for help. They began reading up on experiments such as those conducted by Wolfram Schultz. They asked a Harvard Business School professor to conduct psychological tests of Febreze's ad campaigns. They interviewed customer after customer, looking for something that would give them a clue how to make Febreze a regular part of consumers' lives.

One day, they went to speak with a woman in a suburb near Scottsdale. She was in her forties with four kids. Her house was clean, but not compulsively tidy. To the surprise of the researchers, she loved Febreze.

"I use it every day," she told them.

"You do?" Stimson said. The house didn't seem like the kind of place with smelly problems. There weren't any pets. No one smoked. "How? What smells are you trying to get rid of?"

"I don't really use it for specific smells," the woman said. "I mean, you know, I've got boys. They're going through puberty, and if I don't clean their rooms, it smells like a locker. But I don't really use it that way. I use it for normal cleaning—a couple of sprays when I'm done in a room. It's a nice way to make everything smell good as a final touch."

They asked if they could watch her clean the house. In the bedroom, she made her bed, plumped the pillows, tightened the sheet's corners, and then took a Febreze bottle and sprayed the smoothed comforter. In the living room, she vacuumed, picked up the kids' shoes, straightened the coffee table, and sprayed Febreze on the freshly cleaned carpet. "It's nice, you know?" she said. "Spraying feels like a little mini-celebration when I'm done with a room." At the rate she was using Febreze, Stimson estimated, she would empty a bottle every two weeks.

P&G had collected thousands of hours of videotapes of people cleaning their homes over the years. When the researchers got back to Cincinnati, some of them spent an evening looking through the tapes. The next morning, one of the scientists asked the Febreze team to join him in the conference room. He cued up the tape of one woman—a twenty-six-year-old with three children—making a bed. She smoothed the sheets and adjusted a pillow. Then, she smiled and left the room.

"Did you see that?" the researcher asked excitedly.

He put on another clip. A younger, brunette woman spread out a colorful bedspread, straightened a pillow, and then smiled at her handiwork. "There it is again!" the researcher said. The next clip showed a woman in workout clothes tidying her kitchen and wiping the counter before easing into a relaxing stretch.

The researcher looked at his colleagues.

"Do you see it?" he asked.

"Each of them is doing something relaxing or happy when they finish cleaning," he said. "We can build off that! What if Febreze was something that happened at the *end* of the cleaning routine, rather than the beginning? What if it was the fun part of making something cleaner?"

Stimson's team ran one more test. Previously, the product's advertising had focused on eliminating bad smells. The company printed up new labels that showed open windows and gusts of fresh air. More perfume was added to the recipe, so that instead of merely neutralizing odors, Febreze had its own distinct scent. Television commercials were filmed of women spraying freshly made beds and spritzing just-laundered clothing. The tagline had been "Gets bad smells out of fabrics." It was rewritten as "Cleans life's smells."

Each change was designed to appeal to a specific, daily cue: Cleaning a room. Making a bed. Vacuuming a rug. In each one, Febreze was positioned as the reward: the nice smell that occurs at the end of a cleaning routine. Most important, each ad was calibrated to elicit a craving: that things will smell as nice as they look when the cleaning ritual is done. The irony is that a product manufactured to destroy odors was transformed into the opposite. Instead of eliminating scents on dirty fabrics, it became an air freshener used as the finishing touch, once things are already clean.

When the researchers went back into consumers' homes after the new ads aired and the redesigned bottles were given away, they found that some housewives in the test market had started expecting—craving—the Febreze scent. One woman said that when her bottle ran dry, she squirted diluted perfume on her laundry. "If

I don't smell something nice at the end, it doesn't really seem clean now," she told them.

"The park ranger with the skunk problem sent us in the wrong direction," Stimson told me. "She made us think that Febreze would succeed by providing a solution to a problem. But who wants to admit their house stinks?

"We were looking at it all wrong. No one craves scentlessness. On the other hand, lots of people crave a nice smell after they've spent thirty minutes cleaning."

THE FEBREZE HABIT LOOP

The Febreze relaunch took place in the summer of 1998. Within two months, sales doubled. Within a year, customers had spent more than $230 million on the product. Since then, Febreze has spawned dozens of spin-offs—air fresheners, candles, laundry detergents, and kitchen sprays—that, all told, now account for sales of more than $1 billion per year. Eventually, P&G began mentioning to customers that, in addition to smelling good, Febreze can also kill bad odors.

Stimson was promoted and his team received their bonuses. The formula had worked. They had found simple and obvious cues. They had clearly defined the reward.

But only once they created a sense of craving—the desire to make everything smell as nice as it looked—did Febreze become a hit. That craving is an essential part of the formula for creating new habits that Claude Hopkins, the Pepsodent ad man, never recognized.

V.

In his final years of life, Hopkins took to the lecture circuit. His talks on the "Laws of Scientific Advertising" attracted thousands of people. From stages, he often compared himself to Thomas Edison and George Washington and spun out wild forecasts about the future (flying automobiles featured prominently). But he never mentioned cravings or the neurological roots of the habit loop. After all, it would be another seventy years before the MIT scientists and Wolfram Schultz conducted their experiments.

So how did Hopkins manage to build such a powerful tooth-brushing habit without the benefit of those insights?

Well, it turns out that he actually *did* take advantage of the principles eventually discovered at MIT and inside Schultz's laboratory, even if nobody knew it at the time.

Hopkins's experiences with Pepsodent weren't quite as straight-forward as he portrays them in his memoirs. Though he boasted that he discovered an amazing cue in tooth film, and bragged that he was the first to offer consumers the clear reward of beautiful teeth, it turns out that Hopkins wasn't the originator of those tactics. Not by a long shot. Consider, for instance, some of the advertisements for other toothpastes that filled magazines and newspapers even before Hopkins knew that Pepsodent existed.

"The ingredients of this preparation are especially intended to prevent deposits of *tartar* from accumulating around the necks of the teeth," read an ad for Dr. Sheffield's Crème Dentifrice that pre-dated Pepsodent. "Clean that dirty layer!"

"Your white enamel is only *hidden* by a coating of film," read an advertisement that appeared while Hopkins was looking through his dental textbooks. "Sanitol Tooth Paste quickly restores the original whiteness by removing film."

"The charm of a lovely smile depends upon the beauty of your teeth," proclaimed a third ad. "Beautiful, satin smooth teeth are often the secret of a pretty girl's attractiveness. Use S.S. White Toothpaste!"

Dozens of other advertising men had used the same language as Pepsodent years before Hopkins jumped in the game. All of their ads had promised to remove tooth film and had offered the reward of beautiful, white teeth. None of them had worked.

But once Hopkins launched his campaign, sales of Pepsodent exploded. Why was Pepsodent different?

Because Hopkins's success was driven by the same factors that caused Julio the monkey to touch the lever and housewives to spray Febreze on freshly made beds. Pepsodent created a craving.

Hopkins doesn't spend any of his autobiography discussing the ingredients in Pepsodent, but the recipe listed on the toothpaste's patent application and company records reveals something interesting: Unlike other pastes of the period, Pepsodent contained citric acid, as well as doses of mint oil and other chemicals. Pepsodent's inventor used those ingredients to make the toothpaste taste fresh, but they had another, unanticipated effect as well. They're irritants that create a cool, tingling sensation on the tongue and gums.

After Pepsodent started dominating the marketplace, researchers at competing companies scrambled to figure out why. What they found was that customers said that if they forgot to use Pepsodent, they realized their mistake because they missed that cool, tingling sensation in their mouths. They expected—they *craved*—that slight irritation. If it wasn't there, their mouths didn't feel clean.

Claude Hopkins wasn't selling beautiful teeth. He was selling a sensation. Once people craved that cool tingling—once they equated it with cleanliness—brushing became a habit.

When other companies discovered what Hopkins was really selling, they started imitating him. Within a few decades, almost every toothpaste contained oils and chemicals that caused gums to tingle. Soon, Pepsodent started getting outsold. Even today, almost all

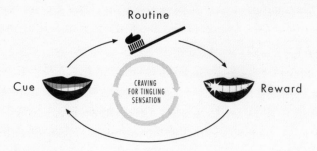

Routine

Cue

CRAVING
FOR TINGLING
SENSATION

Reward

THE REAL PEPSODENT HABIT LOOP

toothpastes contain additives with the sole job of making your mouth tingle after you brush.

"Consumers need some kind of signal that a product is working," Tracy Sinclair, who was a brand manager for Oral-B and Crest Kids Toothpaste, told me. "We can make toothpaste taste like anything—blueberries, green tea—and as long as it has a cool tingle, people feel like their mouth is clean. The tingling doesn't make the toothpaste work any better. It just convinces people it's doing the job."

Anyone can use this basic formula to create habits of her or his own. Want to exercise more? Choose a cue, such as going to the gym as soon as you wake up, and a reward, such as a smoothie after each workout. Then think about that smoothie, or about the endorphin rush you'll feel. Allow yourself to anticipate the reward. Eventually, that craving will make it easier to push through the gym doors every day.

Want to craft a new eating habit? When researchers affiliated with the National Weight Control Registry—a project involving more than six thousand people who have lost more than thirty pounds—looked at the habits of successful dieters, they found that 78 percent of them ate breakfast every morning, a meal cued by a time of day. But most of the successful dieters *also* envisioned a specific reward for sticking with their diet—a bikini they wanted to wear or the sense of pride they felt when they stepped on the scale each day—something they chose carefully and really wanted. They focused on the craving for that reward when temptations arose, cultivated the craving into a mild obsession. And their cravings for that reward, researchers found, crowded out the temptation to drop the diet. The craving drove the habit loop.

For companies, understanding the science of cravings is revolutionary. There are dozens of daily rituals we *ought* to perform each day that never become habits. We should watch our salt and drink more water. We should eat more vegetables and fewer fats. We should take vitamins and apply sunscreen. The facts could not be more clear on this last front: Dabbing a bit of sunscreen on your face each morning significantly lowers the odds of skin cancer. Yet, while everyone brushes their teeth, fewer than 10 percent of Americans apply sunscreen each day. Why?

Because there's no craving that has made sunscreen into a daily habit. Some companies are trying to fix that by giving sunscreens a tingling sensation or something that lets people know they've applied it to their skin. They're hoping it will cue an expectation the same way the craving for a tingling mouth reminds us to brush our teeth. They've already used similar tactics in hundreds of other products.

"Foaming is a huge reward," said Sinclair, the brand manager. "Shampoo doesn't have to foam, but we add foaming chemicals because people expect it each time they wash their hair. Same thing with laundry detergent. And toothpaste—now every company adds sodium laureth sulfate to make toothpaste foam more. There's no cleaning benefit, but people feel better when there's a bunch of suds around their mouth. Once the customer starts expecting that foam, the habit starts growing."

Cravings are what drive habits. And figuring out how to spark a craving makes creating a new habit easier. It's as true now as it was almost a century ago. Every night, millions of people scrub their teeth in order to get a tingling feeling; every morning, millions put on their jogging shoes to capture an endorphin rush they've learned to crave.

And when they get home, after they clean the kitchen or tidy their bedrooms, some of them will spray a bit of Febreze.

3

THE GOLDEN RULE OF HABIT CHANGE
Why Transformation Occurs

I.

The game clock at the far end of the field says there are eight minutes and nineteen seconds left when Tony Dungy, the new head coach of the Tampa Bay Buccaneers—one of the worst teams in the National Football League, not to mention the history of professional football—starts to feel a tiny glimmer of hope.

It's late on a Sunday afternoon, November 17, 1996. The Buccaneers are playing in San Diego against the Chargers, a team that appeared in the Super Bowl the previous year. The Bucs are losing, 17 to 16. They've been losing all game. They've been losing all season. They've been losing all decade. The Buccaneers have not won a game on the West Coast in sixteen years, and many of the team's current players were in grade school the last time the Bucs had a victorious season. So far this year, their record is 2–8. In one of those games, the Detroit Lions—a team so bad it would later be described as putting the "less" in "hopeless"—beat the Bucs 21 to 6, and then, three weeks later, beat them again, 27 to 0. One newspaper columnist has started referring to the Bucs as "America's Orange Door-mat." ESPN is predicting that Dungy, who got his job only in January, could be fired before the year is done.

On the sidelines, however, as Dungy watches his team arrange itself for the next play, it feels like the sun has finally broken through the clouds. He doesn't smile. He never lets his emotions show during a game. But something is taking place on the field, something he's been working toward for years. As the jeers from the hostile crowd of fifty thousand rain down upon him, Tony Dungy sees something that no one else does. He sees proof that his plan is starting to work.

● ● ●

Tony Dungy had waited an eternity for this job. For seventeen years, he prowled the sidelines as an assistant coach, first at the University of Minnesota, then with the Pittsburgh Steelers, then the Kansas City Chiefs, and then back to Minnesota with the Vikings. Four times in the past decade, he had been invited to interview for head coaching positions with NFL teams.

All four times, the interviews hadn't gone well.

Part of the problem was Dungy's coaching philosophy. In his job interviews, he would patiently explain his belief that the key to winning was changing players' habits. He wanted to get players to stop making so many decisions during a game, he said. He wanted them to react automatically, habitually. If he could instill the right habits, his team would win. Period.

"Champions don't do extraordinary things," Dungy would explain. "They do ordinary things, but they do them without thinking, too fast for the other team to react. They follow the habits they've learned."

How, the owners would ask, are you going to create those new habits?

Oh, no, he wasn't going to create *new* habits, Dungy would answer. Players spent their lives building the habits that got them to the NFL. No athlete is going to abandon those patterns simply because some new coach says to.

So rather than creating new habits, Dungy was going to *change* players' old ones. And the secret to changing old habits was using what was already inside players' heads. Habits are a three-step loop—the cue, the routine, and the reward—but Dungy only wanted to attack the middle step, the routine. He knew from experience that it was easier to convince someone to adopt a new behavior if there was something familiar at the beginning and end.

His coaching strategy embodied an axiom, a Golden Rule of habit change that study after study has shown is among the most powerful tools for creating change. Dungy recognized that you can never truly extinguish bad habits.

Rather, to change a habit, you must keep the old cue, and deliver the old reward, but insert a new routine.

That's the rule: If you use the same cue, and provide the same reward, you can shift the routine and change the habit. Almost any behavior can be transformed if the cue and reward stay the same.

The Golden Rule has influenced treatments for alcoholism, obesity, obsessive-compulsive disorders, and hundreds of other destructive behaviors, and understanding it can help anyone change their own habits. (Attempts to give up snacking, for instance, will often fail unless there's a new routine to satisfy old cues and reward urges. A smoker usually can't quit unless she finds some activity to replace cigarettes when her nicotine craving is triggered.)

Four times Dungy explained his habit-based philosophy to team owners. Four times they listened politely, thanked him for his time, and then hired someone else.

Then, in 1996, the woeful Buccaneers called. Dungy flew to Tampa Bay and, once again, laid out his plan for how they could win. The day after the final interview, they offered him the job.

THE GOLDEN RULE OF HABIT CHANGE
You Can't Extinguish a Bad Habit,
You Can Only Change It.

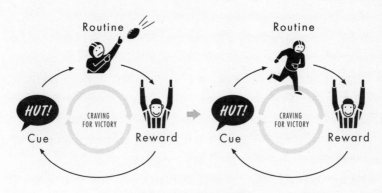

HOW IT WORKS:
USE THE SAME CUE.
PROVIDE THE SAME REWARD.
CHANGE THE ROUTINE.

Dungy's system would eventually turn the Bucs into one of the league's winningest teams. He would become the only coach in NFL history to reach the play-offs in ten consecutive years, the first African American coach to win a Super Bowl, and one of the most respected figures in professional athletics. His coaching techniques would spread throughout the league and all of sports. His approach would help illuminate how to remake the habits in anyone's life.

But all of that would come later. Today, in San Diego, Dungy just wanted to win.

● ● ●

From the sidelines, Dungy looks up at the clock: 8:19 remaining. The Bucs have been behind all game and have squandered opportunity after opportunity, in typical fashion. If their defense doesn't make something happen right now, this game will effectively be over. San Diego has the ball on their own twenty-yard line, and the

Chargers' quarterback, Stan Humphries, is preparing to lead a drive that, he hopes, will put the game away. The play clock begins, and Humphries is poised to take the snap.

But Dungy isn't looking at Humphries. Instead, he's watching his own players align into a formation they have spent months perfecting. Traditionally, football is a game of feints and counterfeints, trick plays and misdirection. Coaches with the thickest playbooks and most complicated schemes usually win. Dungy, however, has taken the opposite approach. He isn't interested in complication or obfuscation. When Dungy's defensive players line up, it is obvious to everyone exactly which play they are going to use.

Dungy has opted for this approach because, in theory, he doesn't need misdirection. He simply needs his team to be faster than everyone else. In football, milliseconds matter. So instead of teaching his players hundreds of formations, he has taught them only a handful, but they have practiced over and over until the behaviors are automatic. When his strategy works, his players can move with a speed that is impossible to overcome.

But only when it works. If his players think too much or hesitate or second-guess their instincts, the system falls apart. And so far, Dungy's players have been a mess.

This time, however, as the Bucs line up on the twenty-yard line, something is different. Take Regan Upshaw, a Buccaneer defensive end who has settled into a three-point stance on the scrimmage line. Instead of looking up and down the line, trying to absorb as much information as possible, Upshaw is looking only at the cues that Dungy taught him to focus on. First, he glances at the outside foot of the opposite lineman (his toes are back, which means he is preparing to step backward and block while the quarterback passes); next, Upshaw looks at the lineman's shoulders (rotated slightly inward), and the space between him and the next player (a fraction narrower than expected).

Upshaw has practiced how to react to each of these cues so many times that, at this point, he doesn't have to think about what to do. He just follows his habits.

San Diego's quarterback approaches the line of scrimmage and glances right, then left, barks the count and takes the ball. He drops back five steps and stands tall, swiveling his head, looking for an open receiver. Three seconds have passed since the play started. The stadium's eyes and the television cameras are on him.

So most observers fail to see what's happening among the Buccaneers. As soon as Humphries took the snap, Upshaw sprang into action. Within the first second of the play, he darted right, across the line of scrimmage, so fast the offensive lineman couldn't block him. Within the next second, Upshaw ran four more paces downfield, his steps a blur. In the next second, Upshaw moved three strides closer to the quarterback, his path impossible for the offensive lineman to predict.

As the play moves into its fourth second, Humphries, the San Diego quarterback, is suddenly exposed. He hesitates, sees Upshaw from the corner of his eye. And that's when Humphries makes his mistake. He starts *thinking*.

Humphries spots a teammate, a rookie tight end named Brian Roche, twenty yards downfield. There's another San Diego receiver much closer, waving his arms, calling for the ball. The short pass is the safe choice. Instead, Humphries, under pressure, performs a split-second analysis, cocks his arm, and heaves to Roche.

That hurried decision is precisely what Dungy was hoping for. As soon as the ball is in the air, a Buccaneer safety named John Lynch starts moving. Lynch's job was straightforward: When the play started, he ran to a particular point on the field and waited for his cue. There's enormous pressure to improvise in this situation. But Dungy has drilled Lynch until his routine is automatic. And as a result, when the ball leaves the quarterback's hands, Lynch is standing ten yards from Roche, waiting.

As the ball spins through the air, Lynch reads his cues—the direction of the quarterback's face mask and hands, the spacing of the receivers—and starts moving before it's clear where the ball will land. Roche, the San Diego receiver, springs forward, but Lynch cuts around him and intercepts the pass. Before Roche can react, Lynch takes off down the field toward the Chargers' end zone. The other Buccaneers are perfectly positioned to clear his route. Lynch runs 10, then 15, then 20, then almost 25 yards before he is finally pushed out of bounds. The entire play has taken less than ten seconds.

Two minutes later, the Bucs score a touchdown, taking the lead for the first time all game. Five minutes later, they kick a field goal. In between, Dungy's defense shuts down each of San Diego's comeback attempts. The Buccaneers win, 25 to 17, one of the biggest upsets of the season.

At the end of the game, Lynch and Dungy exit the field together.

"It feels like something was different out there," Lynch says as they walk into the tunnel.

"We're starting to believe," Dungy replies.

II.

To understand how a coach's focus on changing habits could remake a team, it's necessary to look outside the world of sports. Way outside, to a dingy basement on the Lower East Side of New York City in 1934, where one of the largest and most successful attempts at wide-scale habit change was born.

Sitting in the basement was a thirty-nine-year-old alcoholic named Bill Wilson. Years earlier, Wilson had taken his first drink during officers' training camp in New Bedford, Massachusetts, where he was learning to fire machine guns before getting shipped to France and World War I. Prominent families who lived near the base often invited officers to dinner, and one Sunday night, Wilson attended a party where he was served rarebit and beer. He was twenty-two years old and had never had alcohol before. The only polite thing, it seemed, was to drink the glass served to him. A few weeks later, Wilson was invited to another elegant affair. Men were in tuxedos, women were flirting. A butler came by and put a Bronx cocktail—a combination of gin, dry and sweet vermouth, and orange juice—into Wilson's hand. He took a sip and felt, he later said, as if he had found "the elixir of life."

By the mid-1930s, back from Europe, his marriage falling apart and a fortune from selling stocks vaporized, Wilson was consuming three bottles of booze a day. On a cold November afternoon, while he was sitting in the gloom, an old drinking buddy called. Wilson invited him over and mixed a pitcher of pineapple juice and gin. He poured his friend a glass.

His friend handed it back. He'd been sober for two months, he said.

Wilson was astonished. He started describing his own struggles with alcohol, including the fight he'd gotten into at a country club that had cost him his job. He had tried to quit, he said, but couldn't manage it. He'd been to detox and had taken pills. He'd made promises to his wife and joined abstinence groups. None of it worked. How, Wilson asked, had his friend done it?

"I got religion," the friend said. He talked about hell and temptation, sin and the devil. "Realize you are licked, admit it, and get willing to turn your life over to God."

Wilson thought the guy was nuts. "Last summer an alcoholic crackpot; now, I suspected, a little cracked about religion," he later wrote. When his friend left, Wilson polished off the booze and went to bed.

A month later, in December 1934, Wilson checked into the Charles B. Towns Hospital for Drug and Alcohol Addictions, an upscale Manhattan detox center. A physician started hourly infusions of a hallucinogenic drug called belladonna, then in vogue for the treatment of alcoholism. Wilson floated in and out of consciousness on a bed in a small room.

Then, in an episode that has been described at millions of meetings in cafeterias, union halls, and church basements, Wilson began writhing in agony. For days, he hallucinated. The withdrawal pains made it feel as if insects were crawling across his skin. He was so nauseous he could hardly move, but the pain was too intense to stay still. "If there is a God, let Him show Himself!" Wilson yelled to his empty room. "I am ready to do anything. Anything!" At that moment, he later wrote, a white light filled his room, the pain ceased, and he felt as if he were on a mountaintop, "and that a wind not of air but of spirit was blowing. And then it burst upon me that I was a free man. Slowly the ecstasy subsided. I lay on the bed, but now for a time I was in another world, a new world of consciousness."

Bill Wilson would never have another drink. For the next thirty-six years, until he died of emphysema in 1971, he would devote himself to founding, building, and spreading Alcoholics Anonymous, until it became the largest, most well-known and successful habit-changing organization in the world.

An estimated 2.1 million people seek help from AA each year, and as many as 10 million alcoholics may have achieved sobriety through the group. AA doesn't work for everyone—success rates are difficult to measure, because of participants' anonymity—but millions credit the program with saving their lives. AA's foundational credo, the famous twelve steps, have become cultural lodestones incorporated into treatment programs for overeating, gambling, debt, sex, drugs, hoarding, self-mutilation, smoking, video game addictions, emotional dependency, and dozens of other destructive behaviors. The group's techniques offer, in many respects, one of the most powerful formulas for change.

All of which is somewhat unexpected, because AA has almost no grounding in science or most accepted therapeutic methods.

Alcoholism, of course, is more than a habit. It's a physical addiction with psychological and perhaps genetic roots. What's interesting about AA, however, is that the program doesn't directly attack many of the psychiatric or biochemical issues that researchers say are often at the core of why alcoholics drink. In fact, AA's methods seem to sidestep scientific and medical findings altogether, as well as the types of intervention many psychiatrists say alcoholics really need.*

What AA provides instead is a method for attacking the *habits* that surround alcohol use. AA, in essence, is a giant machine for changing habit loops. And though the habits associated with alcoholism are extreme, the lessons AA provides demonstrate how almost any habit—even the most obstinate—can be changed.

● ● ●

Bill Wilson didn't read academic journals or consult many doctors before founding AA. A few years after he achieved sobriety, he wrote the now-famous twelve steps in a rush one night while sitting in bed. He chose the number twelve because there were twelve apos-

---

* The line separating habits and addictions is often difficult to measure. For instance, the American Society of Addiction Medicine defines addiction as "a primary, chronic disease of brain reward, motivation, memory and related circuitry. . . . Addiction is characterized by impairment in behavioral control, craving, inability to consistently abstain, and diminished relationships."

By that definition, some researchers note, it is difficult to determine why spending fifty dollars a week on cocaine is bad, but fifty dollars a week on coffee is okay. Someone who craves a latte every afternoon may seem clinically addicted to an observer who thinks five dollars for coffee demonstrates an "impairment in behavioral control." Is someone who would prefer running to having breakfast with his kids addicted to exercise?

In general, say many researchers, while addiction is complicated and still poorly understood, many of the behaviors that we associate with it are often driven by habit. Some substances, such as drugs, cigarettes, or alcohol, can create physical dependencies. But these physical cravings often fade quickly after use is discontinued. A physical addiction to nicotine, for instance, lasts only as long as the chemical is in a smoker's bloodstream—about one hundred hours after the last cigarette. Many of the lingering urges that we think of as nicotine's addictive twinges are really behavioral habits asserting themselves—we crave a cigarette at breakfast a month later not because we physically need it, but because we remember so fondly the rush it once provided each morning. Attacking the behaviors we think of as addictions by modifying the habits surrounding them has been shown, in clinical studies, to be one of the most effective modes of treatment. (Though it is worth noting that some chemicals, such as opiates, can cause prolonged physical addictions, and some studies indicate that a small group of people seem predisposed to seek out addictive chemicals, regardless of behavioral interventions. The number of chemicals that cause long-term physical addictions, however, is relatively small, and the number of predisposed addicts is estimated to be much less than the number of alcoholics and addicts seeking help.)

tles. And some aspects of the program are not just unscientific, they can seem downright strange.

Take, for instance, AA's insistence that alcoholics attend "ninety meetings in ninety days"—a stretch of time, it appears, chosen at random. Or the program's intense focus on spirituality, as articulated in step three, which says that alcoholics can achieve sobriety by making "a decision to turn our will and our lives over to the care of God as we understand him." Seven of the twelve steps mention God or spirituality, which seems odd for a program founded by a onetime agnostic who, throughout his life, was openly hostile toward organized religion. AA meetings don't have a prescribed schedule or curriculum. Rather, they usually begin with a member telling his or her story, after which other people can chime in. There are no professionals who guide conversations and few rules about how meetings are supposed to function. In the past five decades, as almost every aspect of psychiatry and addiction research has been revolutionized by discoveries in behavioral sciences, pharmacology, and our understanding of the brain, AA has remained frozen in time.

Because of the program's lack of rigor, academics and researchers have often criticized it. AA's emphasis on spirituality, some claimed, made it more like a cult than a treatment. In the past fifteen years, however, a reevaluation has begun. Researchers now say the program's methods offer valuable lessons. Faculty at Harvard, Yale, the University of Chicago, the University of New Mexico, and dozens of other research centers have found a kind of science within AA that is similar to the one Tony Dungy used on the football field. Their findings endorse the Golden Rule of habit change: AA succeeds because it helps alcoholics use the same cues, and get the same reward, but it shifts the routine.

Researchers say that AA works because the program forces people to identify the cues and rewards that encourage their alcoholic habits, and then helps them find new behaviors. When Claude Hopkins was selling Pepsodent, he found a way to create a new habit by triggering a new craving. But to change an old habit, you must address an old craving. You have to keep the same cues and rewards as before, and feed the craving by inserting a new routine.

Take steps four (to make "a searching and fearless inventory of ourselves") and five (to admit "to God, to ourselves, and to another human being the exact nature of our wrongs").

"It's not obvious from the way they're written, but to complete those steps, someone has to create a list of all the triggers for their alcoholic urges," said J. Scott Tonigan, a researcher at the University of New Mexico who has studied AA for more than a decade. "When you make a self-inventory, you're figuring out all the things that make you drink. And admitting to someone else all the bad things you've done is a pretty good way of figuring out the moments where everything spiraled out of control."

Then, AA asks alcoholics to search for the rewards they get from alcohol. What cravings, the program asks, are driving your habit loop? Often, intoxication itself doesn't make the list. Alcoholics crave a drink because it offers escape, relaxation, companionship, the blunting of anxieties, and an opportunity for emotional release. They might crave a cocktail to forget their worries. But they don't necessarily crave feeling drunk. The physical effects of alcohol are often one of the least rewarding parts of drinking for addicts.

"There is a hedonistic element to alcohol," said Ulf Mueller, a German neurologist who has studied brain activity among alcoholics. "But people also use alcohol because they want to forget something or to satisfy other cravings, and these relief cravings occur in totally different parts of the brain than the craving for physical pleasure."

In order to offer alcoholics the same rewards they get at a bar, AA has built a system of meetings and companionship—the "sponsor" each member works with—that strives to offer as much escape, distraction, and catharsis as a Friday night bender. If someone needs relief, they can get it from talking to their sponsor or attending a group gathering, rather than toasting a drinking buddy.

"AA forces you to create new routines for what to do each night instead of drinking," said Tonigan. "You can relax and talk through your anxieties at the meetings. The triggers and payoffs stay the same, it's just the behavior that changes."

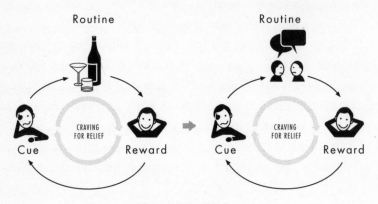

KEEP THE CUE,
PROVIDE THE SAME REWARD,
INSERT A NEW ROUTINE

One particularly dramatic demonstration of how alcoholics' cues and rewards can be transferred to new routines occurred in 2007, when Mueller, the German neurologist, and his colleagues at the University of Magdeburg implanted small electrical devices inside the brains of five alcoholics who had repeatedly tried to give up booze. The alcoholics in the study had each spent at least six months in rehab without success. One of them had been through detox more than sixty times.

The devices implanted in the men's heads were positioned inside their basal ganglia—the same part of the brain where the MIT researchers found the habit loop—and emitted an electrical charge that interrupted the neurological reward that triggers habitual cravings. After the men recovered from the operations, they were exposed to cues that had once triggered alcoholic urges, such as photos of beer or trips to a bar. Normally, it would have been impossible for them to resist a drink. But the devices inside their brains "overrode" each man's neurological cravings. They didn't touch a drop.

"One of them told me the craving disappeared as soon as we turned the electricity on," Mueller said. "Then, we turned it off, and the craving came back immediately."

Eradicating the alcoholics' neurological cravings, however, wasn't enough to stop their drinking habits. Four of them relapsed soon after the surgery, usually after a stressful event. They picked up a bottle because that's how they automatically dealt with anxiety. However, once they learned alternate routines for dealing with stress, the drinking stopped for good. One patient, for instance, attended AA meetings. Others went to therapy. And once they incorporated those new routines for coping with stress and anxiety into their lives, the successes were dramatic. The man who had gone to detox sixty times never had another drink. Two other patients had started drinking at twelve, were alcoholics by eighteen, drank every day, and now have been sober for four years.

Notice how closely this study hews to the Golden Rule of habit change: Even when alcoholics' brains were changed through surgery, it wasn't enough. The old cues and cravings for rewards were still there, waiting to pounce. The alcoholics only permanently changed once they learned new routines that drew on the old triggers and provided a familiar relief. "Some brains are so addicted to alcohol that only surgery can stop it," said Mueller. "But those people also need new ways for dealing with life."

AA provides a similar, though less invasive, system for inserting new routines into old habit loops. As scientists have begun understanding how AA works, they've started applying the program's methods to other habits, such as two-year-olds' tantrums, sex addictions, and even minor behavioral tics. As AA's methods have spread, they've been refined into therapies that can be used to disrupt almost any pattern.

●●●

In the summer of 2006, a twenty-four-year-old graduate student named Mandy walked into the counseling center at Mississippi State University. For most of her life, Mandy had bitten her nails, gnawing them until they bled. Lots of people bite their nails. For chronic nail biters, however, it's a problem of a different scale. Mandy would often bite until her nails pulled away from the skin underneath. Her fingertips were covered with tiny scabs. The end of her fingers had become blunted without nails to protect them and sometimes they tingled or itched, a sign of nerve injury. The biting habit had damaged her social life. She was so embarrassed around her friends that she kept her hands in her pockets and, on dates, would become preoccupied with balling her fingers into fists. She had tried to stop by painting her nails with foul-tasting polishes or promising herself, starting *right now*, that she would muster the willpower to quit. But as soon as she began doing homework or watching television, her fingers ended up in her mouth.

The counseling center referred Mandy to a doctoral psychology student who was studying a treatment known as "habit reversal training." The psychologist was well acquainted with the Golden Rule of habit change. He knew that changing Mandy's nail biting habit required inserting a new routine into her life.

"What do you feel right before you bring your hand up to your mouth to bite your nails?" he asked her.

"There's a little bit of tension in my fingers," Mandy said. "It hurts a little bit here, at the edge of the nail. Sometimes I'll run my thumb along, looking for hangnails, and when I feel something catch, I'll bring it up to my mouth. Then I'll go finger by finger, biting all the rough edges. Once I start, it feels like I have to do all of them."

Asking patients to describe what triggers their habitual behavior is called awareness training, and like AA's insistence on forcing alcoholics to recognize their cues, it's the first step in habit reversal

training. The tension that Mandy felt in her nails cued her nail biting habit.

"Most people's habits have occurred for so long they don't pay
attention to what causes it anymore," said Brad Dufrene, who treated
Mandy. "I've had stutterers come in, and I'll ask them which words
or situations trigger their stuttering, and they won't know because
they stopped noticing so long ago."

Next, the therapist asked Mandy to describe why she bit her nails.
At first, she had trouble coming up with reasons. As they talked,
though, it became clear that she bit when she was bored. The therapist put her in some typical situations, such as watching television
and doing homework, and she started nibbling. When she had
worked through all of the nails, she felt a brief sense of complete-
ness, she said. That was the habit's reward: a physical stimulation
she had come to crave.

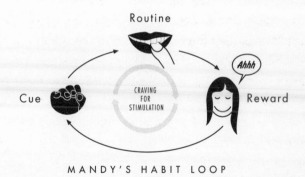

MANDY'S HABIT LOOP

At the end of their first session, the therapist sent Mandy home
with an assignment: Carry around an index card, and each time you
feel the cue—a tension in your fingertips—make a check mark on
the card. She came back a week later with twenty-eight checks. She
was, by that point, acutely aware of the sensations that preceded her
habit. She knew how many times it occurred during class or while
watching television.

Then the therapist taught Mandy what is known as a "competing response." Whenever she felt that tension in her fingertips, he told her, she should immediately put her hands in her pockets or under her legs, or grip a pencil or something else that made it impossible to put her fingers in her mouth. Then Mandy was to search for something that would provide a quick physical stimulation—such as rubbing her arm or rapping her knuckles on a desk—anything that would produce a physical response.

The cues and rewards stayed the same. Only the routine changed.

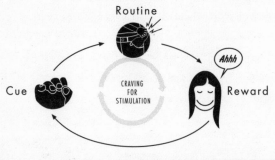

MANDY'S NEW HABIT LOOP

They practiced in the therapist's office for about thirty minutes and Mandy was sent home with a new assignment: Continue with the index card, but make a check when you feel the tension in your fingertips and a hash mark when you successfully override the habit.

A week later, Mandy had bitten her nails only three times and had used the competing response seven times. She rewarded herself with a manicure, but kept using the note cards. After a month, the nail-biting habit was gone. The competing routines had become automatic. One habit had replaced another.

"It seems ridiculously simple, but once you're aware of how your habit works, once you recognize the cues and rewards, you're halfway to changing it," Nathan Azrin, one of the developers of habit reversal training, told me. "It seems like it should be more complex.

The truth is, the brain can be reprogrammed. You just have to be deliberate about it."*

Today, habit reversal therapy is used to treat verbal and physical tics, depression, smoking, gambling problems, anxiety, bedwetting, procrastination, obsessive-compulsive disorders, and other behavioral problems. And its techniques lay bare one of the fundamental principles of habits: Often, we don't really understand the cravings driving our behaviors until we look for them. Mandy never realized that a craving for physical stimulation was causing her nail biting, but once she dissected the habit, it became easy to find a new routine that provided the same reward.

Say you want to stop snacking at work. Is the reward you're seeking to satisfy your hunger? Or is it to interrupt boredom? If you snack for a brief release, you can easily find another routine—such as taking

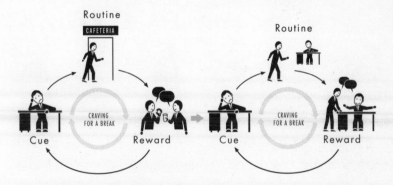

* It is important to note that though the process of habit change is easily described, it does not necessarily follow that it is easily accomplished. It is facile to imply that smoking, alcoholism, overeating, or other ingrained patterns can be upended without real effort. Genuine change requires work and self-understanding of the cravings driving behaviors. Changing any habit requires determination. No one will quit smoking cigarettes simply because they sketch a habit loop.

However, by understanding habits' mechanisms, we gain insights that make new behaviors easier to grasp. Anyone struggling with addiction or destructive behaviors can benefit from help from many quarters, including trained therapists, physicians, social workers, and clergy. Even professionals in those fields, though, agree that most alcoholics, smokers, and other people struggling with problematic behaviors quit on their own, away from formal treatment settings. Much of the time, those changes are accomplished because people examine the cues, cravings, and rewards that drive their behaviors and then find ways to replace their self-destructive routines with healthier alternatives, even if they aren't fully aware of what they are doing at the time. Understanding the cues and cravings driving your habits won't make them suddenly disappear—but it will give you a way to plan how to change the pattern.

a quick walk, or giving yourself three minutes on the Internet—that provides the same interruption without adding to your waistline.

If you want to stop smoking, ask yourself, do you do it because you love nicotine, or because it provides a burst of stimulation, a structure to your day, a way to socialize? If you smoke because you need stimulation, studies indicate that some caffeine in the afternoon can increase the odds you'll quit. More than three dozen studies of former smokers have found that identifying the cues and rewards they associate with cigarettes, and then choosing new routines that provide similar payoffs—a piece of Nicorette, a quick series of push-ups, or simply taking a few minutes to stretch and relax—makes it more likely they will quit.

If you identify the cues and rewards, you can change the routine.

At least, most of the time. For some habits, however, there's one other ingredient that's necessary: belief.

III.

"Here are the six reasons everyone thinks we can't win," Dungy told his Buccaneers after becoming head coach in 1996. It was months before the season started and everyone was sitting in the locker room. Dungy started listing the theories they had all read in the newspapers or heard on the radio: The team's management was messed up. Their new coach was untested. The players were spoiled. The city didn't care. Key players were injured. They didn't have the talent they needed.

"Those are the supposed reasons," Dungy said. "Now here is a fact: Nobody is going to outwork us."

Dungy's strategy, he explained, was to shift the team's behaviors until their performances were automatic. He didn't believe the Buccaneers needed the thickest playbook. He didn't think they had to memorize hundreds of formations. They just had to learn a few key moves and get them right every time.

However, perfection is hard to achieve in football. "Every play in football—every play—someone messes up," said Herm Edwards, one of Dungy's assistant coaches in Tampa Bay. "Most of the time, it's not physical. It's mental." Players mess up when they start thinking too much or second-guessing their plays. What Dungy wanted was to take all that decision making out of their game.

And to do that, he needed them to recognize their existing habits and accept new routines.

He started by watching how his team already played.

"Let's work on the Under Defense," Dungy shouted at a morning practice one day. "Number fifty-five, what's your read?"

"I'm watching the running back and guard," said Derrick Brooks, an outside linebacker.

"What precisely are you *looking* at? Where are your eyes?"

"I'm looking at the movement of the guard," said Brooks. "I'm watching the QB's legs and hips after he gets the ball. And I'm looking for gaps in the line, to see if they're gonna pass and if the QB is going to throw to my side or away."

In football, these visual cues are known as "keys," and they're critical to every play. Dungy's innovation was to use these keys as cues for reworked habits. He knew that, sometimes, Brooks hesitated a moment too long at the start of a play. There were so many things for him to think about—is the guard stepping out of formation? Does the running back's foot indicate he's preparing for a running or passing play?—that sometimes he slowed down.

Dungy's goal was to free Brooks's mind from all that analysis. Like Alcoholics Anonymous, he used the same cues that Brooks was already accustomed to, but gave him different routines that, eventually, occurred automatically.

"I want you to use those same keys," Dungy told Brooks. "But at first, focus only on the running back. That's it. Do it without thinking. Once you're in position, *then* start looking for the QB."

This was a relatively modest shift—Brooks's eyes went to the same cues, but rather than looking multiple places at once, Dungy put them in a sequence and told him, ahead of time, the choice to make when he saw each key. The brilliance of this system was that it removed the need for decision making. It allowed Brooks to move faster, because everything was a reaction—and eventually a habit—rather than a choice.

Dungy gave every player similar instructions, and practiced the formations over and over. It took almost a year for Dungy's habits to take hold. The team lost early, easy games. Sports columnists asked why the Bucs were wasting so much time on psychological quackery.

But slowly, they began to improve. Eventually, the patterns became so familiar to players that they unfolded automatically when the team took the field. In Dungy's second season as coach, the Bucs won their first five games and went to the play-offs for the first time in fifteen years. In 1999, they won the division championship.

Dungy's coaching style started drawing national attention. The sports media fell in love with his soft-spoken demeanor, religious piety, and the importance he placed on balancing work and family. Newspaper stories described how he brought his sons, Eric and Jamie, to the stadium so they could hang out during practice. They did their homework in his office and picked up towels in the locker room. It seemed like, finally, success had arrived.

In 2000, the Bucs made it to the play-offs again, and then again in 2001. Fans now filled the stadium every week. Sportscasters talked about the team as Super Bowl contenders. It was all becoming real.

● ● ●

But even as the Bucs became a powerhouse, a troubling problem emerged. They often played tight, disciplined games. However, during crucial, high-stress moments, everything would fall apart.

In 1999, after racking up six wins in a row at the end of the season, the Bucs blew the conference championship against the St. Louis Rams. In 2000, they were one game away from the Super Bowl when they disintegrated against the Philadelphia Eagles, losing 21 to 3. The next year, the same thing happened again, and the Bucs lost to the Eagles, 31 to 9, blowing their chance of advancing.

"We would practice, and everything would come together and then we'd get to a big game and it was like the training disappeared," Dungy told me. "Afterward, my players would say, 'Well, it was a critical play and I went back to what I knew,' or 'I felt like I had to step it up.' What they were *really* saying was they trusted our system most of the time, but when everything was on the line, that belief broke down."

At the conclusion of the 2001 season, after the Bucs had missed the Super Bowl for the second straight year, the team's general manager asked Dungy to come to his house. He parked near a huge oak tree, walked inside, and thirty seconds later was fired.

The Bucs would go on to win the Super Bowl the next year using Dungy's formations and players, and by relying on the habits he had shaped. He would watch on television as the coach who replaced him lifted up the Lombardi trophy. But by then, he would already be far away.

IV.

About sixty people—soccer moms and lawyers on lunch breaks, old guys with fading tattoos and hipsters in skinny jeans—are sitting in a church and listening to a man with a slight paunch and a tie that complements his pale blue eyes. He looks like a successful politician, with the warm charisma of assured reelection.

"My name is John," he says, "and I'm an alcoholic."

"Hi, John," everyone replies.

"The first time I decided to get help was when my son broke his arm," John says. He's standing behind a podium. "I was having an affair with a woman at work, and she told me that she wanted to end it. So I went to a bar and had two vodkas, and went back to my desk, and at lunch I went to Chili's with a friend, and we each had a few beers, and then at about two o'clock, me and another friend left and found a place with a two-for-one happy hour. It was my day to pick up the kids—my wife didn't know about the affair yet—so I drove to their school and got them, and I was driving home on a street I must have driven a thousand times, and I slammed into a stop sign at the end of the block. Up on the sidewalk and, bam, right into the sign. Sam—that's my boy—hadn't put on his seat belt, so he flew against the windshield and broke his arm. There was blood on the dash where he hit his nose and the windshield was cracked and I was so scared. That's when I decided I needed help.

"So I checked into a clinic and then came out, and everything was pretty good for a while. For about thirteen months, everything was great. I felt like I was in control and I went to meetings every couple of days, but eventually I started thinking, *I'm not such a loser that I need to hang out with a bunch of drunks.* So I stopped going.

"Then my mom got cancer, and she called me at work, almost two years after I got sober. She was driving home from the doctor's office, and she said, 'He told me we can treat it, but it's pretty advanced.' The first thing I did after I hung up is find a bar, and I was pretty much drunk for the next two years until my wife moved out, and I was supposed to pick up my kids again. I was in a really bad place by then. A friend was teaching me to use coke, and every afternoon I would do a line inside my office, and five minutes later I would get that little drip into the back of my throat and do another line.

"Anyways, it was my turn to get the kids. I was on the way to their school and I felt totally fine, like I was on top of everything, and I pulled into an intersection when the light was red and this huge truck slammed into my car. It actually flipped the car on its side. I didn't have a scratch on me. I got out, and started trying to push my car over, because I figured, if I can make it home and leave before the cops arrive, I'll be fine. Of course that didn't work out, and when they arrested me for DUI they showed me how the passenger side of the car was completely crushed in. That's where Sammy usually sat. If he had been there, he would have been killed.

"So I started going to meetings again, and my sponsor told me that it didn't matter if I felt in control. Without a higher power in my life, without admitting my powerlessness, none of it was going to work. I thought that was bull—I'm an atheist. But I knew that if something didn't change, I was going to kill my kids. So I started working at that, working at believing in something bigger than me. And it's working. I don't know if it's God or something else, but there is a power that has helped me stay sober for seven years now and I'm in awe of it. I don't wake up sober every morning—I mean, I haven't had a drink in seven years, but some mornings I wake up feeling like I'm gonna fall down that day. Those days, I look for the higher power, and I call my sponsor, and most of the time we don't talk about drinking. We talk about life and marriage and my job, and by the time I'm ready for a shower, my head is on straight."

The first cracks in the theory that Alcoholics Anonymous succeeded solely by reprogramming participants' habits started appearing a little over a decade ago and were caused by stories from alcoholics like John. Researchers began finding that habit replacement worked pretty well for many people until the stresses of life—such as finding out your mom has cancer, or your marriage is coming apart—got too high, at which point alcoholics often fell off the wagon. Academics asked why, if habit replacement is so effective, it seemed to fail at such critical moments. And as they dug into alcoholics' stories to answer that question, they learned that replacement habits only become durable new behaviors when they are accompanied by something else.

One group of researchers at the Alcohol Research Group in California, for instance, noticed a pattern in interviews. Over and over again, alcoholics said the same thing: Identifying cues and choosing new routines is important, but without another ingredient, the new habits never fully took hold.

The secret, the alcoholics said, was God.

Researchers hated that explanation. God and spirituality are not testable hypotheses. Churches are filled with drunks who continue drinking despite a pious faith. In conversations with addicts, though, spirituality kept coming up again and again. So in 2005, a group of scientists—this time affiliated with UC Berkeley, Brown University, and the National Institutes of Health—began asking alcoholics about all kinds of religious and spiritual topics. Then they looked at the data to see if there was any correlation between religious belief and how long people stayed sober.

A pattern emerged. Alcoholics who practiced the techniques of habit replacement, the data indicated, could often stay sober until there was a stressful event in their lives—at which point, a certain number started drinking again, no matter how many new routines they had embraced.

However, those alcoholics who believed, like John in Brooklyn, that some higher power had entered their lives were more likely to make it through the stressful periods with their sobriety intact.

It wasn't God that mattered, the researchers figured out. It was belief itself that made a difference. Once people learned how to believe in something, that skill started spilling over to other parts of their lives, until they started believing they could change. Belief was the ingredient that made a reworked habit loop into a permanent behavior.

"I wouldn't have said this a year ago—that's how fast our understanding is changing," said Tonigan, the University of New Mexico researcher, "but belief seems critical. You don't have to believe in God, but you do need the capacity to believe that things will get better.

"Even if you give people better habits, it doesn't repair why they started drinking in the first place. Eventually they'll have a bad day, and no new routine is going to make everything seem okay. What can make a difference is *believing* that they can cope with that stress without alcohol."

By putting alcoholics in meetings where belief is a given—where, in fact, belief is an integral part of the twelve steps—AA trains people in how to believe in something until they believe in the program and themselves. It lets people practice believing that things will eventually get better, until things actually do.

"At some point, people in AA look around the room and think, *if it worked for that guy, I guess it can work for me*," said Lee Ann Kaskutas, a senior scientist at the Alcohol Research Group. "There's something really powerful about groups and shared experiences. People might be skeptical about their ability to change if they're by themselves, but a group will convince them to suspend disbelief. A community creates belief."

As John was leaving the AA meeting, I asked him why the program worked now, after it had failed him before. "When I started coming to meetings after the truck accident, someone asked for volunteers to help put away the chairs," he told me. "I raised my hand.

It wasn't a big thing, it took like five minutes, but it felt good to do something that wasn't all about *me*. I think that started me on a different path.

"I wasn't ready to give in to the group the first time, but when I came back, I was ready to start believing in something."

V.

Within a week of Dungy's firing by the Bucs, the owner of the Indianapolis Colts left an impassioned fifteen-minute message on his answering machine. The Colts, despite having one of the NFL's best quarterbacks, Peyton Manning, had just finished a dreadful season. The owner needed help. He was tired of losing, he said. Dungy moved to Indianapolis and became head coach.

He immediately started implementing the same basic game plan: remaking the Colts' routines and teaching players to use old cues to build reworked habits. In his first season, the Colts went 10–6 and qualified for the play-offs. The next season, they went 12–4 and came within one game of the Super Bowl. Dungy's celebrity grew. Newspaper and television profiles appeared around the country. Fans flew in so they could visit the church Dungy attended. His sons became fixtures in the Colts' locker room and on the sidelines. In 2005, Jamie, his eldest boy, graduated from high school and went to college in Florida.

Even as Dungy's successes mounted, however, the same troubling patterns emerged. The Colts would play a season of disciplined, winning football, and then under play-off pressure, choke.

"Belief is the biggest part of success in professional football," Dungy told me. "The team *wanted* to believe, but when things got really tense, they went back to their comfort zones and old habits."

The Colts finished the 2005 regular season with fourteen wins and two losses, the best record in its history.

Then tragedy struck.

Three days before Christmas, Tony Dungy's phone rang in the middle of the night. His wife answered and handed him the receiver, thinking it was one of his players. There was a nurse on the line. Dungy's son Jamie had been brought into the hospital earlier in the evening, she said, with compression injuries on his throat. His girlfriend had found him hanging in his apartment, a belt around his neck. Paramedics had rushed him to the hospital, but efforts at revival were unsuccessful. He was gone.

A chaplain flew to spend Christmas with the family. "Life will never be the same again," the chaplain told them, "but you won't always feel like you do right now."

A few days after the funeral, Dungy returned to the sidelines. He needed something to distract himself, and his wife and team encouraged him to go back to work. "I was overwhelmed by their love and support," he later wrote. "As a group, we had always leaned on each other in difficult times; I needed them now more than ever."

The team lost their first play-off game, concluding their season. But in the aftermath of watching Dungy during this tragedy, "something changed," one of his players from that period told me. "We had seen Coach through this terrible thing and all of us wanted to help him somehow."

It is simplistic, even cavalier, to suggest that a young man's death can have an impact on football games. Dungy has always said that nothing is more important to him than his family. But in the wake of Jamie's passing, as the Colts started preparing for the next season, something shifted, his players say. The team gave in to Dungy's vision of how football should be played in a way they hadn't before. They started to believe.

"I had spent a lot of previous seasons worrying about my contract and salary," said one player who, like others, spoke about that period on the condition of anonymity. "When Coach came back, after the funeral, I wanted to give him everything I could, to take away his hurt. I kind of gave myself to the team."

"Some men like hugging each other," another player told me. "I don't. I haven't hugged my sons in a decade. But after Coach came back, I walked over and I hugged him as long as I could, because I wanted him to know that I was there for him."

After the death of Dungy's son, the team started playing differently. A conviction emerged among players about the strength of Dungy's strategy. In practices and scrimmages leading up to the start of the 2006 season, the Colts played tight, precise football.

"Most football teams aren't really teams. They're just guys who work together," a third player from that period told me. "But we became a *team*. It felt amazing. Coach was the spark, but it was about more than him. After he came back, it felt like we really believed in each other, like we knew how to play together in a way we didn't before."

For the Colts, a belief in their team—in Dungy's tactics and their ability to win—began to emerge out of tragedy. But just as often, a similar belief can emerge without any kind of adversity.

In a 1994 Harvard study that examined people who had radically changed their lives, for instance, researchers found that some people had remade their habits after a personal tragedy, such as a divorce or a life-threatening illness. Others changed after they saw a friend go through something awful, the same way that Dungy's players watched him struggle.

Just as frequently, however, there was no tragedy that preceded people's transformations. Rather, they changed because they were embedded in social groups that made change easier. One woman said her entire life shifted when she signed up for a psychology class and met a wonderful group. "It opened a Pandora's box," the woman told researchers. "I could not tolerate the status quo any longer. I had changed in my core." Another man said that he found new friends among whom he could practice being gregarious. "When I do make the effort to overcome my shyness, I feel that it is not really me acting, that it's someone else," he said. But by practicing with his new group, it stopped feeling like acting. He started to believe he wasn't shy, and then, eventually, he wasn't anymore. When people join groups where change seems possible, the potential for that change to occur becomes more real. For most people who overhaul their lives, there are no seminal moments or life-altering disasters. There are simply communities—sometimes of just one other person—who make change believable. One woman told researchers her life transformed after a day spent cleaning toilets—and after weeks of discussing with the rest of the cleaning crew whether she should leave her husband.

"Change occurs among other people," one of the psychologists involved in the study, Todd Heatherton, told me. "It seems real when we can see it in other people's eyes."

The precise mechanisms of belief are still little understood. No one is certain why a group encountered in a psychology class can convince a woman that everything is different, or why Dungy's team came together after their coach's son passed away. Plenty of people talk to friends about unhappy marriages and never leave their spouses; lots of teams watch their coaches experience adversity and never gel.

But we do know that for habits to permanently change, people must believe that change is feasible. The same process that makes AA so effective—the power of a group to teach individuals how to believe—happens whenever people come together to help one another change. Belief is easier when it occurs within a community.

● ● ●

Ten months after Jamie's death, the 2006 football season began. The Colts played peerless football, winning their first nine games, and finishing the year 12–4. They won their first play-off game, and then beat the Baltimore Ravens for the divisional title. At that point, they were one step away from the Super Bowl, playing for the conference championship—the game that Dungy had lost eight times before.

The matchup occurred on January 21, 2007, against the New England Patriots, the same team that had snuffed out the Colts' Super Bowl aspirations twice.

The Colts started the game strong, but before the first half ended, they began falling apart. Players were afraid of making mistakes or so eager to get past the final Super Bowl hurdle that they lost track of where they were supposed to be focusing. They stopped relying on their habits and started thinking too much. Sloppy tackling led to turnovers. One of Peyton Manning's passes was intercepted and re-turned for a touchdown. Their opponents, the Patriots, pulled ahead 21 to 3. No team in the history of the NFL had ever overcome so big a deficit in a conference championship. Dungy's team, once again, was going to lose.

At halftime, the team filed into the locker room, and Dungy asked everyone to gather around. The noise from the stadium fil-tered through the closed doors, but inside everyone was quiet. Dungy looked at his players.

They had to believe, he said.

"We faced this same situation—against this same team—in 2003," Dungy told them. In that game, they had come within one yard of winning. One yard. "Get your sword ready because this time we're going to win. This is *our* game. It's *our* time."

The Colts came out in the second half and started playing as they had in every preceding game. They stayed focused on their cues and habits. They carefully executed the plays they had spent the past five years practicing until they had become automatic. Their offense, on the opening drive, ground out seventy-six yards over fourteen plays and scored a touchdown. Then, three minutes after taking the next possession, they scored again.

As the fourth quarter wound down, the teams traded points. Dungy's Colts tied the game, but never managed to pull ahead. With

3:49 left in the game, the Patriots scored, putting Dungy's players at a three-point disadvantage, 34 to 31. The Colts got the ball and began driving down the field. They moved seventy yards in nineteen seconds, and crossed into the end zone. For the first time, the Colts had the lead, 38 to 34. There were now sixty seconds left on the clock. If Dungy's team could stop the Patriots from scoring a touchdown, the Colts would win.

Sixty seconds is an eternity in football.

The Patriots' quarterback, Tom Brady, had scored touchdowns in far less time. Sure enough, within seconds of the start of play, Brady moved his team halfway down the field. With seventeen seconds remaining, the Patriots were within striking distance, poised for a final big play that would hand Dungy another defeat and crush, yet again, his team's Super Bowl dreams.

As the Patriots approached the line of scrimmage, the Colts' defense went into their stances. Marlin Jackson, a Colts cornerback, stood ten yards back from the line. He looked at his cues: the width of the gaps between the Patriot linemen and the depth of the running back's stance. Both told him this was going to be a passing play. Tom Brady, the Patriots' quarterback, took the snap and dropped back to pass. Jackson was already moving. Brady cocked his arm and heaved the ball. His intended target was a Patriot receiver twenty-two yards away, wide open, near the middle of the field. If the receiver caught the ball, it was likely he could make it close to the end zone or score a touchdown. The football flew through the air. Jackson, the Colts cornerback, was already running at an angle, following his habits. He rushed past the receiver's right shoulder, cutting in front of him just as the ball arrived. Jackson plucked the ball out of the air for an interception, ran a few more steps and then slid to the ground, hugging the ball to his chest. The whole play had taken less than five seconds. The game was over. Dungy and the Colts had won.

Two weeks later, they won the Super Bowl. There are dozens of reasons that might explain why the Colts finally became champions that year. Maybe they got lucky. Maybe it was just their time. But Dungy's players say it's because they *believed,* and because that belief made everything they had learned—all the routines they had practiced until they became automatic—stick, even at the most stressful moments.

"We're proud to have won this championship for our leader, Coach Dungy," Peyton Manning told the crowd afterward, cradling the Lombardi Trophy.

Dungy turned to his wife. "We did it," he said.

● ● ●

How do habits change?

There is, unfortunately, no specific set of steps guaranteed to work for every person. We know that a habit cannot be eradicated—it must, instead, be replaced. And we know that habits are most malleable when the Golden Rule of habit change is applied: If we keep the same cue and the same reward, a new routine can be inserted.

But that's not enough. For a habit to stay changed, people must believe change is possible. And most often, that belief only emerges with the help of a group.

If you want to quit smoking, figure out a different routine that will satisfy the cravings filled by cigarettes. Then, find a support group, a collection of other former smokers, or a community that will help you believe you can stay away from nicotine, and use that group when you feel you might stumble.

If you want to lose weight, study your habits to determine why you *really* leave your desk for a snack each day, and then find someone else to take a walk with you, to gossip with at their desk rather than in the cafeteria, a group that tracks weight-loss goals together, or someone who also wants to keep a stock of apples, rather than chips, nearby.

The evidence is clear: If you want to change a habit, you must find an alternative routine, and your odds of success go up dramatically when you commit to changing as part of a group. Belief is essential, and it grows out of a communal experience, even if that community is only as large as two people.

We know that change *can* happen. Alcoholics can stop drinking. Smokers can quit puffing. Perennial losers can become champions. You can stop biting your nails or snacking at work, yelling at your kids, staying up all night, or worrying over small concerns. And as scientists have discovered, it's not just individual lives that can shift when habits are tended to. It's also companies, organizations, and communities, as the next chapters explain.

PART
TWO

The Habits of Successful Organizations

4

KEYSTONE HABITS, OR THE BALLAD OF PAUL O'NEILL

Which Habits Matter Most

I.

On a blustery October day in 1987, a herd of prominent Wall Street investors and stock analysts gathered in the ballroom of a posh Manhattan hotel. They were there to meet the new CEO of the Aluminum Company of America—or Alcoa, as it was known—a corporation that, for nearly a century, had manufactured everything from the foil that wraps Hershey's Kisses and the metal in Coca-Cola cans to the bolts that hold satellites together.

Alcoa's founder had invented the process for smelting aluminum a century earlier, and since then the company had become one of the largest on earth. Many of the people in the audience had invested millions of dollars in Alcoa stock and had enjoyed a steady return. In the past year, however, investor grumblings started. Alcoa's management had made misstep after misstep, unwisely trying to expand into new product lines while competitors stole customers and profits away.

So there had been a palpable sense of relief when Alcoa's board announced it was time for new leadership. That relief, though, turned to unease when the choice was announced: the new CEO would be a former government bureaucrat named Paul O'Neill. Many on Wall Street had never heard of him. When Alcoa scheduled this meet and greet at the Manhattan ballroom, every major investor asked for an invitation.

A few minutes before noon, O'Neill took the stage. He was fifty-one years old, trim, and dressed in gray pinstripes and a red power tie. His hair was white and his posture military straight. He bounced up the steps and smiled warmly. He looked dignified, solid, confident. Like a chief executive.

Then he opened his mouth.

"I want to talk to you about worker safety," he said. "Every year, numerous Alcoa workers are injured so badly that they miss a day of work. Our safety record is better than the general American workforce, especially considering that our employees work with metals that are 1500 degrees and machines that can rip a man's arm off. But it's not good enough. I intend to make Alcoa the safest company in America. I intend to go for zero injuries."

The audience was confused. These meetings usually followed a predictable script: A new CEO would start with an introduction, make a faux self-deprecating joke—something about how he slept his way through Harvard Business School—then promise to boost profits and lower costs. Next would come an excoriation of taxes, business regulations, and sometimes, with a fervor that suggested firsthand experience in divorce court, lawyers. Finally, the speech would end with a blizzard of buzzwords—"synergy," "rightsizing," and "co-opetition"—at which point everyone could return to their offices, reassured that capitalism was safe for another day.

O'Neill hadn't said anything about profits. He didn't mention taxes. There was no talk of "using alignment to achieve a win-win synergistic market advantage." For all anyone in the audience knew, given his talk of worker safety, O'Neill might be pro-regulation. Or, worse, a Democrat. It was a terrifying prospect.

"Now, before I go any further," O'Neill said, "I want to point out the safety exits in this room." He gestured to the rear of the ballroom. "There's a couple of doors in the back, and in the unlikely event of a fire or other emergency, you should calmly walk out, go down the stairs to the lobby, and leave the building."

Silence. The only noise was the hum of traffic through the windows. Safety? Fire exits? Was this a joke? One investor in the audience knew that O'Neill had been in Washington, D.C., during the sixties. *Guy must have done a lot of drugs,* he thought.

Eventually, someone raised a hand and asked about inventories in the aerospace division. Another asked about the company's capital ratios.

"I'm not certain you heard me," O'Neill said. "If you want to understand how Alcoa is doing, you need to look at our workplace safety figures. If we bring our injury rates down, it won't be because of cheerleading or the nonsense you sometimes hear from other CEOs. It will be because the individuals at this company have agreed to become part of something important: They've devoted themselves to creating a habit of excellence. Safety will be an indicator that we're making progress in changing our habits across the entire institution. That's how we should be judged."

The investors in the room almost stampeded out the doors when the presentation ended. One jogged to the lobby, found a pay phone, and called his twenty largest clients.

"I said, 'The board put a crazy hippie in charge and he's going to kill the company,'" that investor told me. "I ordered them to sell their stock immediately, before everyone else in the room started calling their clients and telling them the same thing.

"It was literally the worst piece of advice I gave in my entire career."

Within a year of O'Neill's speech, Alcoa's profits would hit a record high. By the time O'Neill retired in 2000, the company's annual net income was five times larger than before he arrived, and its market capitalization had risen by $27 billion. Someone who invested a million dollars in Alcoa on the day O'Neill was hired would have earned another million dollars in dividends while he headed the company, and the value of their stock would be five times bigger when he left.

What's more, all that growth occurred while Alcoa became one of the safest companies in the world. Before O'Neill's arrival, almost every Alcoa plant had at least one accident per week. Once his safety plan was implemented, some facilities would go years without a single employee losing a workday due to an accident. The company's worker injury rate fell to one-twentieth the U.S. average.

So how did O'Neill make one of the largest, stodgiest, and most potentially dangerous companies into a profit machine and a bastion of safety?

By attacking one habit and then watching the changes ripple through the organization.

"I knew I had to transform Alcoa," O'Neill told me. "But you can't *order* people to change. That's not how the brain works. So I decided I was going to start by focusing on one thing. If I could start disrupting the habits around one thing, it would spread throughout the entire company."

O'Neill believed that some habits have the power to start a chain reaction, changing other habits as they move through an organization. Some habits, in other words, matter more than others in remaking businesses and lives. These are "keystone habits," and they can influence how people work, eat, play, live, spend, and communicate. Keystone habits start a process that, over time, transforms everything.

Keystone habits say that success doesn't depend on getting every single thing right, but instead relies on identifying a few key priorities and fashioning them into powerful levers. This book's first section explained how habits work, how they can be created and changed. However, where should a would-be habit master start? Understanding keystone habits holds the answer to that question: The habits that matter most are the ones that, when they start to shift, dislodge and remake other patterns.

Keystone habits explain how Michael Phelps became an Olympic champion and why some college students outperform their peers. They describe why some people, after years of trying, suddenly lose forty pounds while becoming more productive at work and still getting home in time for dinner with their kids. And keystone habits explain how Alcoa became one of the best performing stocks in the Dow Jones index, while also becoming one of the safest places on earth.

● ● ●

When Alcoa first approached O'Neill about becoming CEO, he wasn't sure he wanted the job. He'd already earned plenty of money, and his wife liked Connecticut, where they lived. They didn't know anything about Pittsburgh, where Alcoa was headquartered. But before turning down the offer, O'Neill asked for some time to think it over. To help himself make the decision, he started working on a list of what would be his biggest priorities if he accepted the post.

O'Neill had always been a big believer in lists. Lists were how he organized his life. In college at Fresno State—where he finished his courses in a bit over three years, while also working thirty hours a week—O'Neill had drafted a list of everything he hoped to accomplish during his lifetime, including, near the top, "Make a Difference." After graduating in 1960, at a friend's encouragement, O'Neill picked up an application for a federal internship and, along with three hundred thousand others, took the government employment exam. Three thousand people were chosen for interviews. Three hundred of them were offered jobs. O'Neill was one.

He started as a middle manager at the Veterans Administration and was told to learn about computer systems. All the while, O'Neill kept writing his lists, recording why some projects were more successful than others, which contractors delivered on time and which didn't. He was promoted each year. And as he rose through the VA's ranks, he made a name for himself as someone whose lists always seemed to include a bullet point that got a problem solved.

By the mid-1960s, such skills were in high demand in Washington, D.C. Robert McNamara had recently remade the Pentagon by hiring a crop of young mathematicians, statisticians, and computer programmers. President Johnson wanted some whiz kids of his own. So O'Neill was recruited to what eventually became known as the Office of Management and Budget, one of D.C.'s most powerful agencies. Within a decade, at age thirty-eight, he was promoted to deputy director and was, suddenly, among the most influential people in town.

That's when O'Neill's education in organizational habits really started. One of his first assignments was to create an analytical framework for studying how the government was spending money on health care. He quickly figured out that the government's efforts, which should have been guided by logical rules and deliberate priorities, were instead driven by bizarre institutional processes that, in many ways, operated like habits. Bureaucrats and politicians, rather than making decisions, were responding to cues with automatic routines in order to get rewards such as promotions or reelection. It was the habit loop—spread across thousands of people and billions of dollars.

For instance, after World War II, Congress had created a program to build community hospitals. A quarter century later, it was still chugging along, and so whenever lawmakers allocated new health-care funds, bureaucrats immediately started building. The towns where the new hospitals were located didn't necessarily *need* more patient beds, but that didn't matter. What mattered was erecting a big structure that a politician could point to while stumping for votes.

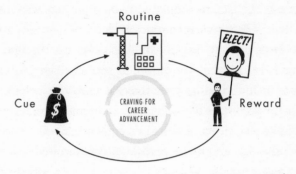

Federal workers would "spend months debating blue or yellow curtains, figuring out if patient rooms should contain one or two televisions, designing nurses' stations, real pointless stuff," O'Neill told me. "Most of the time, no one ever asked if the town wanted a hospital. The bureaucrats had gotten into a habit of solving every medical problem by building something so that a congressman could say, 'Here's what I did!' It didn't make any sense, but everybody did the same thing again and again."

Researchers have found institutional habits in almost every organization or company they've scrutinized. "Individuals have habits; groups have routines," wrote the academic Geoffrey Hodgson, who spent a career examining organizational patterns. "Routines are the organizational analogue of habits."

To O'Neill, these kinds of habits seemed dangerous. "We were basically ceding decision making to a process that occurred without actually thinking," O'Neill said. But at other agencies, where change was in the air, good organizational habits were creating success.

Some departments at NASA, for instance, were overhauling themselves by deliberately instituting organizational routines that encouraged engineers to take more risks. When unmanned rockets exploded on takeoff, department heads would applaud, so that everyone would know their division had tried and failed, but at least they had tried. Eventually, mission control filled with applause every time something expensive blew up. It became an organizational habit. Or take the Environmental Protection Agency, which was created in 1970. The EPA's first administrator, William Ruckelshaus, consciously engineered organizational habits that encouraged his regulators to be aggressive on enforcement. When lawyers asked for permission to file a lawsuit or enforcement action, it went through a process for approval. The default was authorization to go ahead. The message was clear: At the EPA, aggression gets rewarded. By 1975, the EPA was issuing more than fifteen hundred new environmental rules a year.

"Every time I looked at a different part of the government, I found these habits that seemed to explain why things were either succeeding or failing," O'Neill told me. "The best agencies understood the importance of routines. The worst agencies were headed by people who never thought about it, and then wondered why no one followed their orders."

In 1977, after sixteen years in Washington, D.C., O'Neill decided it was time to leave. He was working fifteen hours a day, seven days a week, and his wife was tired of raising four children on her own. O'Neill resigned and landed a job with International Paper, the world's largest pulp and paper company. He eventually became its president.

By then, some of his old government friends were on Alcoa's board. When the company needed a new chief executive, they thought of him, which is how he ended up writing a list of his priorities if he decided to take the job.

At the time, Alcoa was struggling. Critics said the company's workers weren't nimble enough and the quality of its products was poor. But at the top of O'Neill's list he didn't write "quality" or "efficiency" as his biggest priorities. At a company as big and as old as Alcoa, you can't flip a switch and expect everyone to work harder or produce more. The previous CEO had tried to mandate improvements, and fifteen thousand employees had gone on strike. It got so bad they would bring dummies to the parking lots, dress them like managers, and burn them in effigy. "Alcoa was not a happy family," one person from that period told me. "It was more like the Manson family, but with the addition of molten metal."

O'Neill figured his top priority, if he took the job, would have to be something that everybody—unions *and* executives—could agree was important. He needed a focus that would bring people together, that would give him leverage to change how people worked and communicated.

"I went to basics," he told me. "Everyone deserves to leave work as safely as they arrive, right? You shouldn't be scared that feeding your family is going to kill you. That's what I decided to focus on: changing everyone's safety habits."

At the top of O'Neill's list he wrote down "SAFETY" and set an audacious goal: zero injuries. Not zero factory injuries. Zero injuries, period. That would be his commitment no matter how much it cost.

O'Neill decided to take the job.

● ● ●

"I'm really glad to be here," O'Neill told a room full of workers at a smelting plant in Tennessee a few months after he was hired. Not everything had gone smoothly. Wall Street was still panicked. The unions were concerned. Some of Alcoa's vice presidents were miffed at being passed over for the top job. And O'Neill kept talking about worker safety.

"I'm happy to negotiate with you about anything," O'Neill said. He was on a tour of Alcoa's American plants, after which he was going to visit the company's facilities in thirty-one other countries. "But there's one thing I'm never going to negotiate with you, and that's safety. I don't ever want you to say that we haven't taken every step to make sure people don't get hurt. If you want to argue with me about that, you're going to lose."

The brilliance of this approach was that no one, of course, wanted to argue with O'Neill about worker safety. Unions had been fighting for better safety rules for years. Managers didn't want to argue about it, either, since injuries meant lost productivity and low morale.

What most people didn't realize, however, was that O'Neill's plan for getting to zero injuries entailed the most radical realignment in Alcoa's history. The key to protecting Alcoa employees, O'Neill believed, was understanding *why* injuries happened in the first place. And to understand *why* injuries happened, you had to study *how* the manufacturing process was going wrong. To understand *how* things were going wrong, you had to bring in people who could educate workers about quality control and the most efficient work processes, so that it would be easier to do everything right, since correct work is also safer work.

In other words, to protect workers, Alcoa needed to become the best, most streamlined aluminum company on earth.

O'Neill's safety plan, in effect, was modeled on the habit loop. He identified a simple cue: an employee injury. He instituted an automatic routine: Any time someone was injured, the unit president had to report it to O'Neill within twenty-four hours and present a plan for making sure the injury never happened again. And there was a reward: The only people who got promoted were those who embraced the system.

Unit presidents were busy people. To contact O'Neill within twenty-four hours of an injury, they needed to hear about an accident from their vice presidents as soon as it happened. So vice presidents needed to be in constant communication with floor managers. And floor managers needed to get workers to raise warnings as soon as they saw a problem and keep a list of suggestions nearby, so that when the vice president asked for a plan, there was an idea box already full of possibilities. To make all of that happen, each unit had to build new communication systems that made it easier for the lowliest worker to get an idea to the loftiest executive, as fast as possible. Almost everything about the company's rigid hierarchy had to change to accommodate O'Neill's safety program. He was building new corporate habits.

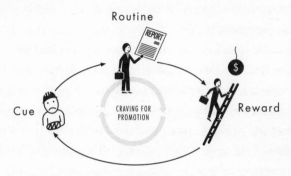

ALCOA'S INSTITUTIONAL HABIT LOOP

As Alcoa's safety patterns shifted, other aspects of the company started changing with startling speed, as well. Rules that unions had spent decades opposing—such as measuring the productivity of individual workers—were suddenly embraced, because such measurements helped everyone figure out when part of the manufacturing process was getting out of whack, posing a safety risk. Policies that managers had long resisted—such as giving workers autonomy to shut down a production line when the pace became overwhelming—were now welcomed, because that was the best way to stop injuries before they occurred. The company shifted so much that some employees found safety habits spilling into other parts of their lives.

"Two or three years ago, I'm in my office, looking at the Ninth Street bridge out the window, and there's some guys working who aren't using correct safety procedures," said Jeff Shockey, Alcoa's current safety director. One of them was standing on top of the bridge's guardrail, while the other held on to his belt. They weren't using safety harnesses or ropes. "They worked for some company that has nothing to do with us, but without thinking about it, I got out of my chair, went down five flights of stairs, walked over the bridge and told these guys, hey, you're risking your life, you have to use your harness and safety gear." The men explained their supervisor had forgotten to bring the equipment. So Shockey called the local Occupational Safety and Health Administration office and turned the supervisor in.

"Another executive told me that one day, he stopped at a street excavation near his house because they didn't have a trench box, and gave everyone a lecture on the importance of proper procedures. It was the weekend, and he stopped his car, with his kids in the back, to lecture city workers about trench safety. That isn't natural, but that's kind of the point. We do this stuff without thinking about it now."

O'Neill never promised that his focus on worker safety would increase Alcoa's profits. However, as his new routines moved through the organization, costs came down, quality went up, and productivity skyrocketed. If molten metal was injuring workers when it splashed, then the pouring system was redesigned, which led to fewer injuries. It also saved money because Alcoa lost less raw materials in spills. If a machine kept breaking down, it was replaced, which meant there was less risk of a broken gear snagging an employee's arm. It also meant higher quality products because, as Alcoa discovered, equipment malfunctions were a chief cause of subpar aluminum.

Researchers have found similar dynamics in dozens of other settings, including individuals' lives.

Take, for instance, studies from the past decade examining the impacts of exercise on daily routines. When people start habitually exercising, even as infrequently as once a week, they start changing other, unrelated patterns in their lives, often unknowingly. Typically, people who exercise start eating better and becoming more productive at work. They smoke less and show more patience with colleagues and family. They use their credit cards less frequently and say they feel less stressed. It's not completely clear why. But for many people, exercise is a keystone habit that triggers widespread change. "Exercise spills over," said James Prochaska, a University of Rhode Island researcher. "There's something about it that makes other good habits easier."

Studies have documented that families who habitually eat dinner together seem to raise children with better homework skills, higher grades, greater emotional control, and more confidence. Making your bed every morning is correlated with better productivity, a greater sense of well-being, and stronger skills at sticking with a budget. It's not that a family meal or a tidy bed *causes* better grades or less frivolous spending. But somehow those initial shifts start chain reactions that help other good habits take hold.

If you focus on changing or cultivating keystone habits, you can cause widespread shifts. However, identifying keystone habits is tricky. To find them, you have to know where to look. Detecting keystone habits means searching out certain characteristics. Keystone habits offer what is known within academic literature as "small wins." They help other habits to flourish by creating new structures, and they establish cultures where change becomes contagious.

But as O'Neill and countless others have found, crossing the gap between understanding those principles and using them requires a bit of ingenuity.

II.

When Michael Phelps's alarm clock went off at 6:30 A.M. on the morning of August 13, 2008, he crawled out of bed in the Olympic Village in Beijing and fell right into his routine.

He pulled on a pair of sweatpants and walked to breakfast. He had already won three gold medals earlier that week—giving him nine in his career—and had two races that day. By 7 A.M., he was in the cafeteria, eating his regular race-day menu of eggs, oatmeal, and four energy shakes, the first of more than six thousand calories he would consume over the next sixteen hours.

Phelps's first race—the 200-meter butterfly, his strongest event—was scheduled for ten o'clock. Two hours before the starting gun fired, he began his usual stretching regime, starting with his arms, then his back, then working down to his ankles, which were so flexible they could extend more than ninety degrees, farther than a ballerina's *en pointe*. At eight-thirty, he slipped into the pool and began his first warm-up lap, 800 meters of mixed styles, followed by 600 meters of kicking, 400 meters pulling a buoy between his legs, 200 meters of stroke drills, and a series of 25-meter sprints to elevate his heart rate. The workout took precisely forty-five minutes.

At nine-fifteen, he exited the pool and started squeezing into his LZR Racer, a bodysuit so tight it required twenty minutes of tugging to put it on. Then he clamped headphones over his ears, cranked up the hip-hop mix he played before every race, and waited.

Phelps had started swimming when he was seven years old to burn off some of the energy that was driving his mom and teachers crazy. When a local swimming coach named Bob Bowman saw Phelps's long torso, big hands, and relatively short legs (which offered less drag in the water), he knew Phelps could become a champion. But Phelps was emotional. He had trouble calming down before races. His parents were divorcing, and he had problems coping with the stress. Bowman purchased a book of relaxation exercises and asked Phelps's mom to read them aloud every night. The book contained a script—"Tighten your right hand into a fist and release it. Imagine the tension melting away"—that tensed and relaxed each part of Phelps's body before he fell asleep.

Bowman believed that for swimmers, the key to victory was creating the right routines. Phelps, Bowman knew, had a perfect physique for the pool. That said, everyone who eventually competes at the Olympics has perfect musculature. Bowman could also see that Phelps, even at a young age, had a capacity for obsessiveness that made him an ideal athlete. Then again, all elite performers are obsessives.

What Bowman could give Phelps, however—what would set him apart from other competitors—were habits that would make him the strongest mental swimmer in the pool. He didn't need to control every aspect of Phelps's life. All he needed to do was target a few specific habits that had nothing to do with swimming and everything to do with creating the right mind-set. He designed a series of behaviors that Phelps could use to become calm and focused before each race, to find those tiny advantages that, in a sport where victory can come in milliseconds, would make all the difference.

When Phelps was a teenager, for instance, at the end of each practice, Bowman would tell him to go home and "watch the videotape. Watch it before you go to sleep and when you wake up."

The videotape wasn't real. Rather, it was a mental visualization of the perfect race. Each night before falling asleep and each morning after waking up, Phelps would imagine himself jumping off the blocks and, in slow motion, swimming flawlessly. He would visualize his strokes, the walls of the pool, his turns, and the finish. He would imagine the wake behind his body, the water dripping off his lips as his mouth cleared the surface, what it would feel like to rip off his cap at the end. He would lie in bed with his eyes shut and watch the entire competition, the smallest details, again and again, until he knew each second by heart.

During practices, when Bowman ordered Phelps to swim at race speed, he would shout, "Put in the videotape!" and Phelps would push himself, as hard as he could. It almost felt anticlimactic as he cut through the water. He had done this so many times in his head that, by now, it felt rote. But it worked. He got faster and faster. Eventually, all Bowman had to do before a race was whisper, "Get the videotape ready," and Phelps would settle down and crush the competition.

And once Bowman established a few core routines in Phelps's life, all the other habits—his diet and practice schedules, the stretching and sleep routines—seemed to fall into place on their own. At the core of why those habits were so effective, why they acted as keystone habits, was something known within academic literature as a "small win."

● ● ●

Small wins are exactly what they sound like, and are part of how keystone habits create widespread changes. A huge body of research has shown that small wins have enormous power, an influence disproportionate to the accomplishments of the victories themselves. "Small wins are a steady application of a small advantage," one Cornell professor wrote in 1984. "Once a small win has been accomplished, forces are set in motion that favor another small win." Small wins fuel transformative changes by leveraging tiny advantages into patterns that convince people that bigger achievements are within reach.

For example, when gay rights organizations started campaigning against homophobia in the late 1960s, their initial efforts yielded only a string of failures. They pushed to repeal laws used to prosecute gays and were roundly defeated in state legislatures. Teachers tried to create curriculums to counsel gay teens, and were fired for suggesting that homosexuality should be embraced. It seemed like the gay community's larger goals—ending discrimination and police harassment, convincing the American Psychiatric Association to stop defining homosexuality as a mental disease—were out of reach.

Then, in the early 1970s, the American Library Association's Task

Force on Gay Liberation decided to focus on one modest goal: convincing the Library of Congress to reclassify books about the gay liberation movement from HQ 71–471 ("Abnormal Sexual Relations, Including Sexual Crimes") to another, less pejorative category.

In 1972, after receiving a letter requesting the reclassification, the Library of Congress agreed to make the shift, reclassifying books into a newly created category, HQ 76.5 ("Homosexuality, Lesbianism—Gay Liberation Movement, Homophile Movement"). It was a minor tweak of an old institutional habit regarding how books were shelved, but the effect was electrifying. News of the new policy spread across the nation. Gay rights organizations, citing the victory, started fund-raising drives. Within a few years, openly gay politicians were running for political office in California, New York, Massachusetts, and Oregon, many of them citing the Library of Congress's decision as inspiration. In 1973, the American Psychiatric Association, after years of internal debate, rewrote the definition of homosexuality so it was no longer a mental illness—paving the way for the passage of state laws that made it illegal to discriminate against people because of their sexual orientation.

And it all began with one small win.

"Small wins do not combine in a neat, linear, serial form, with each step being a demonstrable step closer to some predetermined goal," wrote Karl Weick, a prominent organizational psychologist. "More common is the circumstance where small wins are scattered . . . like miniature experiments that test implicit theories about resistance and opportunity and uncover both resources and barriers that were invisible before the situation was stirred up."

Which is precisely what happened with Michael Phelps. When Bob Bowman started working with Phelps and his mother on the keystone habits of visualization and relaxation, neither Bowman nor Phelps had any idea what they were doing. "We'd experiment, try different things until we found stuff that worked," Bowman told me. "Eventually we figured out it was best to concentrate on these tiny moments of success and build them into mental triggers. We worked them into a routine. There's a series of things we do before every race that are designed to give Michael a sense of building victory.

"If you were to ask Michael what's going on in his head before competition, he would say he's not really thinking about anything. He's just following the program. But that's not right. It's more like his habits have taken over. When the race arrives, he's more than halfway through his plan and he's been victorious at every step. All the stretches went like he planned. The warm-up laps were just like he visualized. His headphones are playing exactly what he expected. The actual race is just another step in a pattern that started earlier that day and has been nothing but victories. Winning is a natural extension."

Back in Beijing, it was 9:56 A.M.—four minutes before the race's start—and Phelps stood behind his starting block, bouncing slightly on his toes. When the announcer said his name, Phelps stepped onto the block, as he always did before a race, and then stepped down, as he always did. He swung his arms three times, as he had before every race since he was twelve years old. He stepped up on the blocks again, got into his stance, and, when the gun sounded, leapt.

Phelps knew that something was wrong as soon as he hit the water. There was moisture inside his goggles. He couldn't tell if they were leaking from the top or bottom, but as he broke the water's surface and began swimming, he hoped the leak wouldn't become too bad.

By the second turn, however, everything was getting blurry. As he approached the third turn and final lap, the cups of his goggles were completely filled. Phelps couldn't see anything. Not the line along the pool's bottom, not the black T marking the approaching wall. He couldn't see how many strokes were left. For most swimmers, losing your sight in the middle of an Olympic final would be cause for panic.

Phelps was calm.

Everything else that day had gone according to plan. The leaking goggles were a minor deviation, but one for which he was prepared. Bowman had once made Phelps swim in a Michigan pool in the dark, believing that he needed to be ready for any surprise. Some of the videotapes in Phelps's mind had featured problems like this. He had mentally rehearsed how he would respond to a goggle failure. As he started his last lap, Phelps estimated how many strokes the final push would require—nineteen or twenty, maybe twenty-one—and started counting. He felt totally relaxed as he swam at full strength. Midway through the lap he began to increase his effort, a final eruption that had become one of his main techniques in overwhelming opponents. At eighteen strokes, he started anticipating the wall. He could hear the crowd roaring, but since he was blind, he had no idea if they were cheering for him or someone else. Nineteen strokes, then twenty. It felt like he needed one more. That's what the videotape in his head said. He made a twenty-first, huge stroke, glided with his arm outstretched, and touched the wall. He had timed it perfectly. When he ripped off his goggles and looked up at the scoreboard, it said "WR"—world record—next to his name. He'd won another gold.

After the race, a reporter asked what it had felt like to swim blind.

"It felt like I imagined it would," Phelps said. It was one additional victory in a lifetime full of small wins.

● ● ●

Six months after Paul O'Neill became CEO of Alcoa, he got a telephone call in the middle of the night. A plant manager in Arizona was on the line, panicked, talking about how an extrusion press had stopped operating and one of the workers—a young man who had joined the company a few weeks earlier, eager for the job because it offered health care for his pregnant wife—had tried a repair. He had jumped over a yellow safety wall surrounding the press and walked across the pit. There was a piece of aluminum jammed into the hinge on a swinging six-foot arm. The young man pulled on the aluminum scrap, removing it. The machine was fixed. Behind him, the arm restarted its arc, swinging toward his head. When it hit, the arm crushed his skull. He was killed instantly.

Fourteen hours later, O'Neill ordered all the plant's executives—as well as Alcoa's top officers in Pittsburgh—into an emergency meeting. For much of the day, they painstakingly re-created the accident with diagrams and by watching videotapes again and again. They identified dozens of errors that had contributed to the death, including two managers who had seen the man jump over the barrier but failed to stop him; a training program that hadn't emphasized to the man that he wouldn't be blamed for a breakdown; lack of instructions that he should find a manager before attempting a repair; and the absence of sensors to automatically shut down the machine when someone stepped into the pit.

"We killed this man," a grim-faced O'Neill told the group. "It's my failure of leadership. I caused his death. And it's the failure of all of you in the chain of command."

The executives in the room were taken aback. Sure, a tragic accident had occurred, but tragic accidents were part of life at Alcoa. It was a huge company with employees who handled red-hot metal and dangerous machines. "Paul had come in as an outsider, and there was a lot of skepticism when he talked about safety," said Bill O'Rourke, a top executive. "We figured it would last a few weeks, and then he would start focusing on something else. But that meeting really shook everyone up. He was serious about this stuff, serious enough that he would stay up nights worrying about some employee he'd never met. That's when things started to change."

Within a week of that meeting, all the safety railings at Alcoa's plants were repainted bright yellow, and new policies were written up. Managers told employees not to be afraid to suggest proactive maintenance, and rules were clarified so that no one would attempt unsafe repairs. The newfound vigilance resulted in a short-term, noticeable decline in the injury rate. Alcoa experienced a small win.

Then O'Neill pounced.

"I want to congratulate everyone for bringing down the number of accidents, even just for two weeks," he wrote in a memo that made its way through the entire company. "We shouldn't celebrate because we've followed the rules, or brought down a number. We should celebrate because we are saving lives."

Workers made copies of the note and taped it to their lockers. Someone painted a mural of O'Neill on one of the walls of a smelting plant with a quote from the memo inscribed underneath. Just as Michael Phelps's routines had nothing to do with swimming and everything to do with his success, so O'Neill's efforts began snowballing into changes that were unrelated to safety, but transformative nonetheless.

"I said to the hourly workers, 'If your management doesn't follow up on safety issues, then call me at home, here's my number,'" O'Neill told me. "Workers started calling, but they didn't want to talk about accidents. They wanted to talk about all these other great ideas."

The Alcoa plant that manufactured aluminum siding for houses, for instance, had been struggling for years because executives would try to anticipate popular colors and inevitably guess wrong. They would pay consultants millions of dollars to choose shades of paint and six months later, the warehouse would be overflowing with "sunburst yellow" and out of suddenly in-demand "hunter green." One day, a low-level employee made a suggestion that quickly worked its way to the general manager: If they grouped all the painting machines together, they could switch out the pigments faster and become more nimble in responding to shifts in customer demand. Within a year, profits on aluminum siding doubled.

The small wins that started with O'Neill's focus on safety created a climate in which all kinds of new ideas bubbled up.

"It turns out this guy had been suggesting this painting idea for a decade, but hadn't told anyone in management," an Alcoa executive told me. "Then he figures, since we keep on asking for safety recommendations, why not tell them about this *other* idea? It was like he gave us the winning lottery numbers."

III.

When a young Paul O'Neill was working for the government and creating a framework for analyzing federal spending on health care, one of the foremost issues concerning officials was infant mortality. The United States, at the time, was one of the wealthiest countries on earth. Yet it had a higher infant mortality rate than most of Europe and some parts of South America. Rural areas, in particular, saw a staggering number of babies die before their first birthdays.

O'Neill was tasked with figuring out why. He asked other federal agencies to start analyzing infant mortality data, and each time someone came back with an answer, he'd ask another question, trying to get deeper, to understand the problem's root causes. Whenever someone came into O'Neill's office with some discovery, O'Neill would start interrogating them with new inquiries. He drove people crazy with his never-ending push to learn more, to understand what was *really* going on. ("I love Paul O'Neill, but you could not pay me enough to work for him again," one official told me. "The man has never encountered an answer he can't turn into another twenty hours of work.")

Some research, for instance, suggested that the biggest cause of infant deaths was premature births. And the reason babies were born too early was that mothers suffered from malnourishment during pregnancy. So to lower infant mortality, improve mothers' diets. Simple, right? But to stop malnourishment, women had to improve their diets *before* they became pregnant. Which meant the government had to start educating women about nutrition before they became sexually active. Which meant officials had to create nutrition curriculums inside high schools.

However, when O'Neill began asking about how to create those curriculums, he discovered that many high school teachers in rural areas didn't know enough basic biology to teach nutrition. So the government had to remake how teachers were getting educated in college, and give them a stronger grounding in biology so they could eventually teach nutrition to teenage girls, so those teenagers would eat better before they started having sex, and, eventually, be sufficiently nourished when they had children.

Poor teacher training, the officials working with O'Neill finally figured out, was a root cause of high infant mortality. If you asked doctors or public health officials for a plan to fight infant deaths, none of them would have suggested changing how teachers are trained. They wouldn't have known there was a link. However, by teaching college students about biology, you made it possible for them to eventually pass on that knowledge to teenagers, who started eating healthier, and years later give birth to stronger babies. Today, the U.S. infant mortality rate is 68 percent lower than when O'Neill started the job.

O'Neill's experiences with infant mortality illustrate the second way that keystone habits encourage change: by creating structures that help other habits to flourish. In the case of premature deaths, changing collegiate curriculums for teachers started a chain reaction that eventually trickled down to how girls were educated in rural areas, and whether they were sufficiently nourished when they became pregnant. And O'Neill's habit of constantly pushing other bureaucrats to continue researching until they found a problem's root causes overhauled how the government thought about problems like infant mortality.

The same thing can happen in people's lives. For example, until about twenty years ago, conventional wisdom held that the best way for people to lose weight was to radically alter their lives. Doctors would give obese patients strict diets and tell them to join a gym, attend regular counseling sessions—sometimes as often as every day—and shift their daily routines by walking up stairs, for instance, instead of taking the elevator. Only by completely shaking up someone's life, the thinking went, could their bad habits be reformed.

But when researchers studied the effectiveness of these methods over prolonged periods, they discovered they were failures. Patients would use the stairs for a few weeks, but by the end of the month, it was too much hassle. They began diets and joined gyms, but after the initial burst of enthusiasm wore off, they slid back into their old eating and TV-watching habits. Piling on so much change at once made it impossible for any of it to stick.

Then, in 2009 a group of researchers funded by the National Institutes of Health published a study of a different approach to weight loss. They had assembled a group of sixteen hundred obese people and asked them to concentrate on writing down everything they ate at least one day per week.

It was hard at first. The subjects forgot to carry their food journals, or would snack and not note it. Slowly, however, people started recording their meals once a week—and sometimes, more often. Many participants started keeping a daily food log. Eventually, it became a habit. Then, something unexpected happened. The participants started looking at their entries and finding patterns they didn't know existed. Some noticed they always seemed to snack at about 10 A.M., so they began keeping an apple or banana on their desks for midmorning munchies. Others started using their journals to plan future menus, and when dinner rolled around, they ate the healthy meal they had written down, rather than junk food from the fridge.

The researchers hadn't suggested any of these behaviors. They had simply asked everyone to write down what they ate once a week. But this keystone habit—food journaling—created a structure that helped other habits to flourish. Six months into the study, people who kept daily food records had lost twice as much weight as everyone else.

"After a while, the journal got inside my head," one person told me. "I started thinking about meals differently. It gave me a system for thinking about food without becoming depressed."

Something similar happened at Alcoa after O'Neill took over. Just as food journals provided a structure for other habits to flourish, O'Neill's safety habits created an atmosphere in which other behaviors emerged. Early on, O'Neill took the unusual step of ordering Alcoa's offices around the world to link up in an electronic network. This was in the early 1980s, when large, international networks weren't usually connected to people's desktop computers. O'Neill justified his order by arguing that it was essential to create a real-time safety data system that managers could use to share suggestions. As a result, Alcoa developed one of the first genuinely worldwide corporate email systems.

O'Neill logged on every morning and sent messages to make sure everyone else was logged on as well. At first, people used the network primarily to discuss safety issues. Then, as email habits became more ingrained and comfortable, they started posting information on all kinds of other topics, such as local market conditions, sales quotas, and business problems. High-ranking executives were required to send in a report every Friday, which anyone in the company could read. A manager in Brazil used the network to send a colleague in New York data on changes in the price of steel. The New Yorker took that information and turned a quick profit for the company on Wall Street. Pretty soon, everyone was using the system to communicate about everything. "I would send in my accident report, and I knew everyone else read it, so I figured, why not send pricing information, or intelligence on other companies?" one manager told me. "It was like we had discovered a secret weapon. The competition couldn't figure out how we were doing it."

When the Web blossomed, Alcoa was perfectly positioned to take advantage. O'Neill's keystone habit—worker safety—had created a platform that encouraged another practice—email—years ahead of competitors.

● ● ●

By 1996, Paul O'Neill had been at Alcoa for almost a decade. His leadership had been studied by the Harvard Business School and the Kennedy School of Government. He was regularly mentioned as a potential commerce secretary or secretary of defense. His employees and the unions gave him high marks. Under his watch, Alcoa's stock price had risen more than 200 percent. He was, at last, a universally acknowledged success.

In May of that year, at a shareholder meeting in downtown Pittsburgh, a Benedictine nun stood up during the question-and-answer session and accused O'Neill of lying. Sister Mary Margaret represented a social advocacy group concerned about wages and conditions inside an Alcoa plant in Ciudad Acuña, Mexico. She said that while O'Neill extolled Alcoa's safety measures, workers in Mexico were becoming sick because of dangerous fumes.

"It's untrue," O'Neill told the room. On his laptop, he pulled up the safety records from the Mexican plant. "See?" he said, showing the room its high scores on safety, environmental compliance, and employee satisfaction surveys. The executive in charge of the facility, Robert Barton, was one of Alcoa's most senior managers. He had been with the company for decades and was responsible for some of their largest partnerships. The nun said that the audience shouldn't trust O'Neill. She sat down.

After the meeting, O'Neill asked her to come to his office. The nun's religious order owned fifty Alcoa shares, and for months they had been asking for a shareholder vote on a resolution to review the company's Mexican operations. O'Neill asked Sister Mary if she had been to any of the plants herself. No, she told him. To be safe, O'Neill asked the company's head of human resources and general counsel to fly to Mexico to see what was going on.

When the executives arrived, they poked through the Acuña plant's records, and found reports of an incident that had never been sent to headquarters. A few months earlier, there had been a buildup of fumes within a building. It was a relatively minor event. The plant's executive, Barton, had installed ventilators to remove the gases. The people who had become ill had fully recovered within a day or two.

But Barton had never reported the illnesses.

When the executives returned to Pittsburgh and presented their findings, O'Neill had a question.

"Did Bob Barton *know* that people had gotten sick?"

"We didn't meet with him," they answered. "But, yeah, it's pretty clear he knew."

Two days later, Barton was fired.

The exit shocked outsiders. Barton had been mentioned in articles as one of the company's most valuable executives. His departure was a blow to important joint ventures.

Within Alcoa, however, no one was surprised. It was seen as an inevitable extension of the culture that O'Neill had built.

"Barton fired himself," one of his colleagues told me. "There wasn't even a choice there."

This is the final way that keystone habits encourage widespread change: by creating cultures where new values become ingrained. Keystone habits make tough choices—such as firing a top executive—easier, because when that person violates the culture, it's clear they have to go. Sometimes these cultures manifest themselves in special vocabularies, the use of which becomes, itself, a habit that defines an organization. At Alcoa, for instance, there were "Core Programs" and "Safety Philosophies," phrases that acted like suitcases, containing whole conversations about priorities, goals, and ways of thinking.

"It might have been hard at another company to fire someone who had been there so long," O'Neill told me. "It wasn't hard for me. It was clear what our values dictated. He got fired because he didn't report the incident, and so no one else had the opportunity to learn from it. Not sharing an opportunity to learn is a cardinal sin."

Cultures grow out of the keystone habits in every organization, whether leaders are aware of them or not. For instance, when researchers studied an incoming class of cadets at West Point, they measured their grade point averages, physical aptitude, military abilities, and self-discipline. When they correlated those factors with whether students dropped out or graduated, however, they found that all of them mattered less than a factor researchers referred to as "grit," which they defined as the tendency to work "strenuously toward challenges, maintaining effort and interest over years despite failure, adversity, and plateaus in progress."

What's most interesting about grit is how it emerges. It grows out of a culture that cadets create for themselves, and that culture often emerges because of keystone habits they adopt at West Point. "There's so much about this school that's hard," one cadet told me. "They call the first summer 'Beast Barracks,' because they want to grind you down. Tons of people quit before the school year starts.

"But I found this group of guys in the first couple of days here, and we started this thing where, every morning, we get together to make sure everyone is feeling strong. I go to them if I'm feeling worried or down, and I know they'll pump me back up. There's only nine of us, and we call ourselves the musketeers. Without them, I don't think I would have lasted a month here."

Cadets who are successful at West Point arrive at the school armed with habits of mental and physical discipline. Those assets, however, only carry you so far. To succeed, they need a keystone habit that creates a culture—such as a daily gathering of like-minded friends—to help find the strength to overcome obstacles. Keystone habits transform us by creating cultures that make clear the values that, in the heat of a difficult decision or a moment of uncertainty, we might otherwise forget.

● ● ●

In 2000, O'Neill retired from Alcoa, and at the request of the newly elected president George W. Bush, became secretary of the treasury.* He left that post two years later, and today spends most of his time teaching hospitals how to focus on worker safety and keystone habits that can lower medical error rates, as well as serving on various corporate boards.

Companies and organizations across America, in the meantime, have embraced the idea of using keystone habits to remake workplaces. At IBM, for instance, Lou Gerstner rebuilt the firm by initially concentrating on one keystone habit: IBM's research and selling routines. At the consulting firm McKinsey & Company, a culture of continuous improvement is created through a keystone habit of wide-ranging internal critiques that are at the core of every assignment. Within Goldman Sachs, a keystone habit of risk assessment undergirds every decision.

And at Alcoa, O'Neill's legacy lives on. Even in his absence, the injury rate has continued to decline. In 2010, 82 percent of Alcoa locations didn't lose one employee day due to injury, close to an all-time high. On average, workers are more likely to get injured at a software company, animating cartoons for movie studios, or doing taxes as an accountant than handling molten aluminum at Alcoa.

"When I was made a plant manager," said Jeff Shockey, the Alcoa

* O'Neill's tenure at Treasury was not as successful as his career at Alcoa. Almost immediately after taking office he began focusing on a couple of key issues, including worker safety, job creation, executive accountability, and fighting African poverty, among other initiatives.

However, O'Neill's politics did not line up with those of President Bush, and he launched an internal fight opposing Bush's proposed tax cuts. He was asked to resign at the end of 2002. "What I thought was the right thing for economic policy was the opposite of what the White House wanted," O'Neill told me. "That's not good for a treasury secretary, so I got fired."

executive, "the first day I pulled into the parking lot I saw all these parking spaces near the front doors with people's titles on them. The head guy for this or that. People who were important got the best parking spots. The first thing I did was tell a maintenance manager to paint over all the titles. I wanted whoever got to work earliest to get the best spot. Everyone understood the message: Every person matters. It was an extension of what Paul was doing around worker safety. It electrified the plant. Pretty soon, everyone was getting to work earlier each day."

5

STARBUCKS AND THE HABIT OF SUCCESS

When Willpower Becomes Automatic

I.

The first time Travis Leach saw his father overdose, he was nine years old. His family had just moved into a small apartment at the end of an alleyway, the latest in a seemingly endless series of relocations that had most recently caused them to abandon their previous home in the middle of the night, throwing everything they owned into black garbage bags after receiving an eviction notice. Too many people coming and going too late at night, the landlord said. Too much noise.

Sometimes, at his old house, Travis would come home from school and find the rooms neatly cleaned, leftovers meticulously wrapped in the fridge and packets of hot sauce and ketchup in Tupperware containers. He knew this meant his parents had temporarily abandoned heroin for crank and spent the day in a cleaning frenzy. Those usually ended badly. Travis felt safer when the house was messy and his parents were on the couch, their eyes half-lidded, watching cartoons. There is no chaos at the end of a heroin fog.

Travis's father was a gentle man who loved to cook and, except for a stint in the navy, spent his entire life within a few miles of his parents in Lodi, California. Travis's mother, by the time everyone moved into the alleyway apartment, was in prison for heroin possession and prostitution. His parents were, essentially, functional addicts and the family maintained a veneer of normalcy. They went camping every summer and on most Friday nights attended his sister and brother's softball games. When Travis was four years old, he went to Disneyland with his dad and was photographed for the first time in his life, by a Disney employee. The family camera had been sold to a pawn shop years before.

On the morning of the overdose, Travis and his brother were playing in the living room on top of blankets they laid out on the floor each night for sleeping. Travis's father was getting ready to make pancakes when he stepped into the bathroom. He was carrying the tube sock that contained his needle, spoon, lighter, and cotton swabs. A few moments later, he came out, opened the refrigerator to get the eggs, and crashed to the floor. When the kids ran around the corner, their father was convulsing, his face turning blue.

Travis's siblings had seen an overdose before and knew the drill. His brother rolled him onto his side. His sister opened his mouth to make sure he wouldn't choke on his tongue, and told Travis to run next door, ask to use the neighbor's phone, and dial 911.

"My name is Travis, my dad is passed out, and we don't know what happened. He's not breathing," Travis lied to the police operator. Even at nine years old, he knew why his father was unconscious. He didn't want to say it in front of the neighbor. Three years earlier, one of his dad's friends had died in their basement after shooting up. When the paramedics had taken the body away, neighbors gawked at Travis and his sister while they held the door open for the gurney. One of the neighbors had a cousin whose son was in his class, and soon everyone in school had known.

After hanging up the phone, Travis walked to the end of the alleyway and waited for the ambulance. His father was treated at the hospital that morning, charged at the police station in the afternoon, and home again by dinnertime. He made spaghetti. Travis turned ten a few weeks later.

● ● ●

When Travis was sixteen, he dropped out of high school. "I was tired of being called a faggot," he said, "tired of people following me home and throwing things at me. Everything seemed really overwhelming. It was easier to quit and go somewhere else." He moved two hours south, to Fresno, and got a job at a car wash. He was fired for insubordination. He got jobs at McDonald's and Hollywood Video, but when customers were rude—"I wanted *ranch dressing, you moron!*"—he would lose control.

"Get out of my drive-through!" he shouted at one woman, throwing the chicken nuggets at her car before his manager pulled him inside.

Sometimes he'd get so upset that he would start crying in the middle of a shift. He was often late, or he'd take a day off for no reason. In the morning, he would yell at his reflection in the mirror, order himself to be better, to suck it up. But he couldn't get along with people, and he wasn't strong enough to weather the steady drip of criticisms and indignities. When the line at his register would get too long and the manager would shout at him, Travis's hands would start shaking and he'd feel like he couldn't catch his breath. He wondered if this is what his parents felt like, so defenseless against life, when they started using drugs.

One day, a regular customer at Hollywood Video who'd gotten to know Travis a little bit suggested he think about working at Starbucks. "We're opening a new store on Fort Washington, and I'm going to be an assistant manager," the man said. "You should apply." A month later, Travis was a barista on the morning shift.

That was six years ago. Today, at twenty-five, Travis is the manager of two Starbucks where he oversees forty employees and is responsible for revenues exceeding $2 million per year. His salary is $44,000 and he has a 401(k) and no debt. He's never late to work. He does not get upset on the job. When one of his employees started crying after a customer screamed at her, Travis took her aside.

"Your apron is a shield," he told her. "Nothing anyone says will ever hurt you. You will always be as strong as you want to be."

He picked up that lecture in one of his Starbucks training courses, an education program that began on his first day and continues throughout an employee's career. The program is sufficiently structured that he can earn college credits by completing the modules. The training has, Travis says, changed his life. Starbucks has taught him how to live, how to focus, how to get to work on time, and how to master his emotions. Most crucially, it has taught him willpower.

"Starbucks is the most important thing that has ever happened to me," he told me. "I owe everything to this company."

● ● ●

For Travis and thousands of others, Starbucks—like a handful of other companies—has succeeded in teaching the kind of life skills that schools, families, and communities have failed to provide. With more than 137,000 current employees and more than one million alumni, Starbucks is now, in a sense, one of the nation's largest educators. All of those employees, in their first year alone, spent at least fifty hours in Starbucks classrooms, and dozens more at home with Starbucks' workbooks and talking to the Starbucks mentors assigned to them.

At the core of that education is an intense focus on an all-important habit: willpower. Dozens of studies show that willpower is the single most important keystone habit for individual success. In a 2005 study, for instance, researchers from the University of Pennsylvania analyzed 164 eighth-grade students, measuring their IQs and other factors, including how much willpower the students demonstrated, as measured by tests of their self-discipline.

Students who exerted high levels of willpower were more likely to earn higher grades in their classes and gain admission into more selective schools. They had fewer absences and spent less time watching television and more hours on homework. "Highly self-disciplined adolescents outperformed their more impulsive peers on every academic-performance variable," the researchers wrote. "Self-discipline predicted academic performance more robustly than did IQ. Self-discipline also predicted which students would improve their grades over the course of the school year, whereas IQ did not. . . . Self-discipline has a bigger effect on academic performance than does intellectual talent."

And the best way to strengthen willpower and give students a leg up, studies indicate, is to make it into a habit. "Sometimes it looks like people with great self-control aren't working hard—but that's because they've made it automatic," Angela Duckworth, one of the University of Pennsylvania researchers told me. "Their willpower occurs without them having to think about it."

For Starbucks, willpower is more than an academic curiosity. When the company began plotting its massive growth strategy in the late 1990s, executives recognized that success required cultivating an environment that justified paying four dollars for a fancy cup of coffee. The company needed to train its employees to deliver a bit of joy alongside lattes and scones. So early on, Starbucks started researching how they could teach employees to regulate their emotions and marshal their self-discipline to deliver a burst of pep with every serving. Unless baristas are trained to put aside their personal problems, the emotions of some employees will inevitably spill into how they treat customers. However, if a worker knows how to remain focused and disciplined, even at the end of an eight-hour shift, they'll deliver the higher class of fast food service that Starbucks customers expect.

The company spent millions of dollars developing curriculums to train employees on self-discipline. Executives wrote workbooks that, in effect, serve as guides to how to make willpower a habit in workers' lives. Those curriculums are, in part, why Starbucks has grown from a sleepy Seattle company into a behemoth with more than seventeen thousand stores and revenues of more than $10 billion a year.

So how does Starbucks do it? How do they take people like Travis—the son of drug addicts and a high school dropout who couldn't muster enough self-control to hold down a job at McDonald's—and teach him to oversee dozens of employees and tens of thousands of dollars in revenue each month? What, precisely, did Travis learn?

II.

Everyone who walked into the room where the experiment was being conducted at Case Western Reserve University agreed on one thing: The cookies smelled delicious. They had just come out of the oven and were piled in a bowl, oozing with chocolate chips. On the table next to the cookies was a bowl of radishes. All day long, hungry students walked in, sat in front of the two foods, and submitted, unknowingly, to a test of their willpower that would upend our understanding of how self-discipline works.

At the time, there was relatively little academic scrutiny into willpower. Psychologists considered such subjects to be aspects of something they called "self-regulation," but it wasn't a field that inspired great curiosity. There was one famous experiment, conducted in the 1960s, in which scientists at Stanford had tested the willpower of a group of four-year-olds. The kids were brought into a room and presented with a selection of treats, including marshmallows. They were offered a deal: They could eat one marshmallow right away, or, if they waited a few minutes, they could have two marshmallows. Then the researcher left the room. Some kids gave in to temptation and ate the marshmallow as soon as the adult left. About 30 percent managed to ignore their urges, and doubled their treats when the researcher came back fifteen minutes later. Scientists, who were watching everything from behind a two-way mirror, kept careful track of which kids had enough self-control to earn the second marshmallow.

Years later, they tracked down many of the study's participants. By now, they were in high school. The researchers asked about their grades and SAT scores, ability to maintain friendships, and their capacity to "cope with important problems." They discovered that the four-year-olds who could delay gratification the longest ended up with the best grades and with SAT scores 210 points higher, on average, than everyone else. They were more popular and did fewer drugs. If you knew how to avoid the temptation of a marshmallow as a preschooler, it seemed, you also knew how to get yourself to class on time and finish your homework once you got older, as well as how to make friends and resist peer pressure. It was as if the marshmallow-ignoring kids had self-regulatory skills that gave them an advantage throughout their lives.

Scientists began conducting related experiments, trying to figure out how to help kids increase their self-regulatory skills. They learned that teaching them simple tricks—such as distracting themselves by drawing a picture, or imagining a frame around the marshmallow, so it seemed more like a photo and less like a real

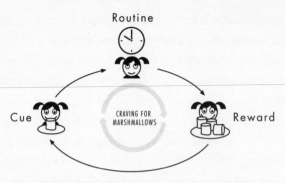

WHEN KIDS LEARN HABITS FOR
DELAYING THEIR CRAVINGS...

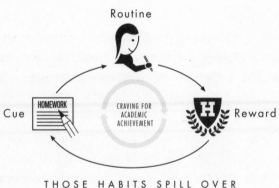

THOSE HABITS SPILL OVER
TO OTHER PARTS OF LIFE

temptation—helped them learn self-control. By the 1980s, a theory emerged that became generally accepted: Willpower is a learnable skill, something that can be taught the same way kids learn to do math and say "thank you." But funding for these inquiries was scarce. The topic of willpower wasn't in vogue. Many of the Stanford scientists moved on to other areas of research.

However, when a group of psychology PhD candidates at Case Western—including one named Mark Muraven—discovered those studies in the mid-nineties, they started asking questions the previous research didn't seem to answer. To Muraven, this model of willpower-as-skill wasn't a satisfying explanation. A skill, after all, is something that remains constant from day to day. If you have the skill to make an omelet on Wednesday, you'll still know how to make it on Friday.

In Muraven's experience, though, it felt like he forgot how to exert willpower all the time. Some evenings he would come home from work and have no problem going for a jog. Other days, he couldn't do anything besides lie on the couch and watch television. It was as if his brain—or, at least, that part of his brain responsible for making him exercise—had forgotten how to summon the will-power to push him out the door. Some days, he ate healthily. Other days, when he was tired, he raided the vending machines and stuffed himself with candy and chips.

If willpower is a skill, Muraven wondered, then why doesn't it remain constant from day to day? He suspected there was more to willpower than the earlier experiments had revealed. But how do you test that in a laboratory?

● ● ●

Muraven's solution was the lab containing one bowl of freshly baked cookies and one bowl of radishes. The room was essentially a closet with a two-way mirror, outfitted with a table, a wooden chair, a hand bell, and a toaster oven. Sixty-seven undergraduates were recruited and told to skip a meal. One by one, the undergrads sat in front of the two bowls.

"The point of this experiment is to test taste perceptions," a researcher told each student, which was untrue. The point was to force students—but only *some* students—to exert their willpower. To that end, half the undergraduates were instructed to eat the cookies and ignore the radishes; the other half were told to eat the radishes and ignore the cookies. Muraven's theory was that ignoring cookies is hard—it takes willpower. Ignoring radishes, on the other hand, hardly requires any effort at all.

"Remember," the researcher said, "eat only the food that has been assigned to you." Then she left the room.

Once the students were alone, they started munching. The cookie eaters were in heaven. The radish eaters were in agony. They were miserable forcing themselves to ignore the warm cookies. Through the two-way mirror, the researchers watched one of the radish eaters pick up a cookie, smell it longingly, and then put it back in the bowl. Another grabbed a few cookies, put them down, and then licked melted chocolate off his fingers.

After five minutes, the researcher reentered the room. By Muraven's estimation, the radish eaters' willpower had been thoroughly taxed by eating the bitter vegetable and ignoring the treats; the cookie eaters had hardly used any of their self-discipline.

"We need to wait about fifteen minutes for the sensory memory of the food you ate to fade," the researcher told each participant. To pass the time, she asked them to complete a puzzle. It looked fairly simple: trace a geometric pattern without lifting your pencil from the page or going over the same line twice. If you want to quit, the researcher said, ring the bell. She implied the puzzle wouldn't take long.

In truth, the puzzle was impossible to solve.

This puzzle wasn't a way to pass time; it was the most important part of the experiment. It took enormous willpower to keep working on the puzzle, particularly when each attempt failed. The scientists wondered, would the students who had already expended their willpower by ignoring the cookies give up on the puzzle faster? In other words, was willpower a finite resource?

From behind their two-way mirror, the researchers watched. The cookie eaters, with their unused reservoirs of self-discipline, started working on the puzzle. In general, they looked relaxed. One of them tried a straightforward approach, hit a roadblock, and then started again. And again. And again. Some worked for over half an hour before the researcher told them to stop. On average, the cookie eaters spent almost nineteen minutes apiece trying to solve the puzzle before they rang the bell.

The radish eaters, with their depleted willpower, acted completely different. They muttered as they worked. They got frustrated. One complained that the whole experiment was a waste of time. Some of them put their heads on the table and closed their eyes. One snapped at the researcher when she came back in. On average, the radish eaters worked for only about eight minutes, 60 percent less time than the cookie eaters, before quitting. When the researcher asked afterward how they felt, one of the radish eaters said he was "sick of this dumb experiment."

"By making people use a little bit of their willpower to ignore cookies, we had put them into a state where they were willing to quit much faster," Muraven told me. "There's been more than two hundred studies on this idea since then, and they've all found the same thing. Willpower isn't just a skill. It's a muscle, like the muscles in your arms or legs, and it gets tired as it works harder, so there's less power left over for other things."

Researchers have built on this finding to explain all sorts of phenomena. Some have suggested it helps clarify why otherwise successful people succumb to extramarital affairs (which are most likely to start late at night after a long day of using willpower at work) or why good physicians make dumb mistakes (which most often occur after a doctor has finished a long, complicated task that requires intense focus). "If you want to do something that requires willpower—like going for a run after work—you have to conserve your willpower muscle during the day," Muraven told me. "If you use it up too early on tedious tasks like writing emails or filling out complicated and boring expense forms, all the strength will be gone by the time you get home."

● ● ●

But how far does this analogy extend? Will exercising willpower muscles make them stronger the same way using dumbbells strengthen biceps?

In 2006, two Australian researchers—Megan Oaten and Ken Cheng—tried to answer that question by creating a willpower workout. They enrolled two dozen people between the ages of eighteen and fifty in a physical exercise program and, over two months, put them through an increasing number of weight lifting, resistance training, and aerobic routines. Week after week, people forced themselves to exercise more frequently, using more and more willpower each time they hit the gym.

After two months, the researchers scrutinized the rest of the participants' lives to see if increased willpower at the gym resulted in greater willpower at home. Before the experiment began, most of the subjects were self-professed couch potatoes. Now, of course, they were in better physical shape. But they were also healthier in other parts of their lives, as well. The more time they spent at the gym, the fewer cigarettes they smoked and the less alcohol, caffeine, and junk food they consumed. They were spending more hours on homework and fewer watching TV. They were less depressed.

Maybe, Oaten and Cheng wondered, those results had nothing to do with willpower. What if exercise just makes people happier and less hungry for fast food?

So they designed another experiment. This time, they signed up twenty-nine people for a four-month money management program. They set savings goals and asked participants to deny themselves luxuries, such as meals at restaurants or movies. Participants were asked to keep detailed logs of everything they bought, which was annoying at first, but eventually people worked up the self-discipline to jot down every purchase.

People's finances improved as they progressed through the program. More surprising, they also smoked fewer cigarettes and drank less alcohol and caffeine—on average, two fewer cups of coffee, two fewer beers, and, among smokers, fifteen fewer cigarettes each day. They ate less junk food and were more productive at work and school. It was like the exercise study: As people strengthened their willpower muscles in one part of their lives—in the gym, or a money management program—that strength spilled over into what they ate or how hard they worked. Once willpower became stronger, it touched everything.

Oaten and Cheng did one more experiment. They enrolled forty-five students in an academic improvement program that focused on creating study habits. Predictably, participants' learning skills improved. And the students also smoked less, drank less, watched less television, exercised more, and ate healthier, even though all those things were never mentioned in the academic program. Again, as their willpower muscles strengthened, good habits seemed to spill over into other parts of their lives.

"When you learn to force yourself to go to the gym or start your homework or eat a salad instead of a hamburger, part of what's happening is that you're changing how you think," said Todd Heatherton, a researcher at Dartmouth who has worked on willpower studies. "People get better at regulating their impulses. They learn how to distract themselves from temptations. And once you've gotten into that willpower groove, your brain is practiced at helping you focus on a goal."

There are now hundreds of researchers, at nearly every major university, studying willpower. Public and charter schools in Philadelphia, Seattle, New York, and elsewhere have started incorporating willpower-strengthening lessons into curriculums. At KIPP, or the "Knowledge Is Power Program"—a collection of charter schools serving low-income students across the nation—teaching self-control is part of the schools' philosophy. (A KIPP school in Philadelphia gave students shirts proclaiming "Don't Eat the Marshmallow.") Many of these schools have dramatically raised students' test scores.

"That's why signing kids up for piano lessons or sports is so important. It has nothing to do with creating a good musician or a five-year-old soccer star," said Heatherton. "When you learn to force yourself to practice for an hour or run fifteen laps, you start building self-regulatory strength. A five-year-old who can follow the ball for ten minutes becomes a sixth grader who can start his homework on time."

As research on willpower has become a hot topic in scientific journals and newspaper articles, it has started to trickle into corporate America. Firms such as Starbucks—and the Gap, Walmart, restaurants, or any other business that relies on entry-level workers—all face a common problem: No matter how much their employees *want* to do a great job, many will fail because they lack self-discipline. They show up late. They snap at rude customers. They get distracted or drawn into workplace dramas. They quit for no reason.

"For a lot of employees, Starbucks is their first professional experience," said Christine Deputy, who helped oversee the company's training programs for more than a decade. "If your parents or teachers have been telling you what to do your entire life, and suddenly customers are yelling and your boss is too busy to give you guidance, it can be really overwhelming. A lot of people can't make the transition. So we try to figure out how to give our employees the self-discipline they didn't learn in high school."

But when companies like Starbucks tried to apply the willpower lessons from the radish-and-cookie studies to the workplace, they encountered difficulties. They sponsored weight-loss classes and offered employees free gym memberships, hoping the benefits would spill over to how they served coffee. Attendance was spotty. It was hard to sit through a class or hit the gym after a full day at work, employees complained. "If someone has trouble with self-discipline at work, they're probably also going to have trouble attending a program designed to strengthen their self-discipline *after* work," Muraven said.

But Starbucks was determined to solve this problem. By 2007, during the height of its expansion, the company was opening seven new stores every day and hiring as many as fifteen hundred employees each week. Training them to excel at customer service—to show up on time and not get angry at patrons and serve everyone with a smile while remembering customers' orders and, if possible, their names— was essential. People expect an expensive latte delivered with a bit of sparkle. "We're not in the coffee business serving people," Howard Behar, the former president of Starbucks, told me. "We're in the people business serving coffee. Our entire business model is based on fantastic customer service. Without that, we're toast."

The solution, Starbucks discovered, was turning self-discipline into an organizational habit.

III.

In 1992, a British psychologist walked into two of Scotland's busiest orthopedic hospitals and recruited five-dozen patients for an experiment she hoped would explain how to boost the willpower of people exceptionally resistant to change.

The patients, on average, were sixty-eight years old. Most of them earned less than $10,000 a year and didn't have more than a high school degree. All of them had recently undergone hip or knee replacement surgeries, but because they were relatively poor and uneducated, many had waited years for their operations. They were retirees, elderly mechanics, and store clerks. They were in life's final chapters, and most had no desire to pick up a new book.

Recovering from a hip or knee surgery is incredibly arduous. The operation involves severing joint muscles and sawing through bones. While recovering, the smallest movements—shifting in bed or flexing a joint—can be excruciating. However, it is essential that patients begin exercising almost as soon as they wake from surgery. They must begin moving their legs and hips before the muscles and skin have healed, or scar tissue will clog the joint, destroying its flexibility. In addition, if patients don't start exercising, they risk developing blood clots. But the agony is so extreme that it's not unusual for people to skip out on rehab sessions. Patients, particularly elderly ones, often refuse to comply with doctors' orders.

The Scottish study's participants were the types of people most likely to fail at rehabilitation. The scientist conducting the experiment wanted to see if it was possible to help them harness their willpower. She gave each patient a booklet after their surgeries that detailed their rehab schedule, and in the back were thirteen additional pages—one for each week—with blank spaces and instructions: "My goals for this week are _____ ? Write down exactly what you are going to do. For example, if you are going to go for a walk this week, write down where and when you are going to walk." She asked patients to fill in each of those pages with specific plans. Then she compared the recoveries of those who wrote out goals with those of patients who had received the same booklets, but didn't write anything.

It seems absurd to think that giving people a few pieces of blank paper might make a difference in how quickly they recover from surgery. But when the researcher visited the patients three months later, she found a striking difference between the two groups. The patients who had written plans in their booklets had started walking almost twice as fast as the ones who had not. They had started getting in and out of their chairs, unassisted, almost three times as fast. They were putting on their shoes, doing the laundry, and making themselves meals quicker than the patients who hadn't scribbled out goals ahead of time.

The psychologist wanted to understand why. She examined the booklets, and discovered that most of the blank pages had been filled in with specific, detailed plans about the most mundane aspects of recovery. One patient, for example, had written, "I will walk to the bus stop tomorrow to meet my wife from work," and then noted what time he would leave, the route he would walk, what he would wear, which coat he would bring if it was raining, and what pills he would take if the pain became too much. Another patient, in a similar study, wrote a series of very specific schedules regarding the exercises he would do each time he went to the bathroom. A third wrote a minute-by-minute itinerary for walking around the block.

As the psychologist scrutinized the booklets, she saw that many of the plans had something in common: They focused on how patients would handle a specific moment of anticipated pain. The man who exercised on the way to the bathroom, for instance, knew that each time he stood up from the couch, the ache was excruciating. So he wrote out a plan for dealing with it: Automatically take the first step, right away, so he wouldn't be tempted to sit down again. The patient who met his wife at the bus stop dreaded the afternoons, because that stroll was the longest and most painful each day. So he detailed every obstacle he might confront, and came up with a solution ahead of time.

Put another way, the patients' plans were built around inflection points when they knew their pain—and thus the temptation to quit—would be strongest. The patients were telling themselves how they were going to make it over the hump.

Each of them, intuitively, employed the same rules that Claude Hopkins had used to sell Pepsodent. They identified simple cues

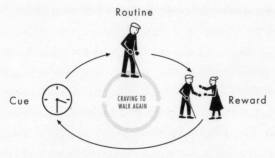

Cue — Routine — CRAVING TO WALK AGAIN — Reward

PATIENTS DESIGNED WILLPOWER
HABITS TO HELP THEM OVERCOME
PAINFUL INFLECTION POINTS

and obvious rewards. The man who met his wife at the bus stop, for instance, identified an easy cue—*It's 3:30, she's on her way home!*—and he clearly defined his reward—*Honey, I'm here!* When the temptation to give up halfway through the walk appeared, the patient could ignore it because he had crafted self-discipline into a habit.

There's no reason why the other patients—the ones who didn't write out recovery plans—couldn't have behaved the same way. All the patients had been exposed to the same admonitions and warnings at the hospital. They all knew exercise was essential for their recovery. They all spent weeks in rehab.

But the patients who didn't write out any plans were at a significant disadvantage, because they never thought ahead about how to deal with painful inflection points. They never deliberately designed willpower habits. Even if they intended to walk around the block, their resolve abandoned them when they confronted the agony of the first few steps.

● ● ●

When Starbucks's attempts at boosting workers' willpower through gym memberships and diet workshops faltered, executives decided they needed to take a new approach. They started by looking more closely at what was actually happening inside their stores. They saw that, like the Scottish patients, their workers were failing when they ran up against inflection points. What they needed were institutional habits that made it easier to muster their self-discipline.

Executives determined that, in some ways, they had been thinking about willpower all wrong. Employees with willpower lapses, it turned out, had no difficulty doing their jobs most of the time. On the average day, a willpower-challenged worker was no different from anyone else. But sometimes, particularly when faced with unexpected stresses or uncertainties, those employees would snap and their self-control would evaporate. A customer might begin yelling, for instance, and a normally calm employee would lose her composure. An impatient crowd might overwhelm a barista, and suddenly he was on the edge of tears.

What employees really needed were clear instructions about how to deal with inflection points—something similar to the Scottish patients' booklets: a routine for employees to follow when their willpower muscles went limp. So the company developed new training materials that spelled out routines for employees to use when they hit rough patches. The manuals taught workers how to respond to specific cues, such as a screaming customer or a long line at a cash register. Managers drilled employees, role-playing with them until the responses became automatic. The company identified specific rewards—a grateful customer, praise from a manager—that employees could look to as evidence of a job well done.

Starbucks taught their employees how to handle moments of adversity by giving them willpower habit loops.

When Travis started at Starbucks, for instance, his manager introduced him to the habits right away. "One of the hardest things about this job is dealing with an angry customer," Travis's manager told him. "When someone comes up and starts yelling at you because they got the wrong drink, what's your first reaction?"

"I don't know," Travis said. "I guess I feel kind of scared. Or angry."

"That's natural," his manager said. "But our job is to provide the best customer service, even when the pressure's on." The manager flipped open the Starbucks manual, and showed Travis a page that was largely blank. At the top, it read, "When a customer is unhappy, my plan is to . . ."

"This workbook is for you to imagine unpleasant situations, and write out a plan for responding," the manager said. "One of the systems we use is called the *LATTE* method. We *Listen* to the customer, *Acknowledge* their complaint, *Take action* by solving the problem, *Thank* them, and then *Explain* why the problem occurred.

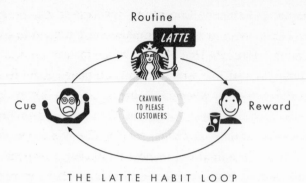

THE LATTE HABIT LOOP

"Why don't you take a few minutes, and write out a plan for dealing with an angry customer. Use the *LATTE* method. Then we can role-play a little bit."

Starbucks has dozens of routines that employees are taught to use during stressful inflection points. There's the *What What Why* system of giving criticism and the *Connect, Discover, and Respond* system for taking orders when things become hectic. There are learned habits to help baristas tell the difference between patrons who just want their coffee ("A hurried customer speaks with a sense of urgency and may seem impatient or look at their watch") and those who need a bit more coddling ("A regular customer knows other baristas by name and normally orders the same beverage each day"). Throughout the training manuals are dozens of blank pages where employees can write out plans that anticipate how they will surmount inflection points. Then they practice those plans, again and again, until they become automatic.

This is how willpower becomes a habit: by choosing a certain behavior ahead of time, and then following that routine when an inflection point arrives. When the Scottish patients filled out their booklets, or Travis studied the *LATTE* method, they decided ahead of time how to react to a cue—a painful muscle or an angry customer. When the cue arrived, the routine occurred.

Starbucks isn't the only company to use such training methods.

For instance, at Deloitte Consulting, the largest tax and financial services company in the world, employees are trained in a curriculum named "Moments That Matter," which focuses on dealing with inflection points such as when a client complains about fees, when a colleague is fired, or when a Deloitte consultant has made a mistake. For each of those moments, there are preprogrammed routines—*Get Curious, Say What No One Else Will, Apply the 5/5/5 Rule*—that guide employees in how they should respond. At the Container Store, employees receive more than 185 hours of training in their first year alone. They are taught to recognize inflection points such as an angry coworker or an overwhelmed customer, and habits, such as routines for calming shoppers or defusing a confrontation. When a customer comes in who seems overwhelmed, for example, an employee immediately asks them to visualize the space in their home they are hoping to organize, and describe how they'll feel when everything is in its place. "We've had customers come up to us and say, 'This is better than a visit to my shrink,'" the company's CEO told a reporter.

IV.

Howard Schultz, the man who built Starbucks into a colossus, isn't so different from Travis in some ways. He grew up in a public housing project in Brooklyn, sharing a two-bedroom apartment with his parents and two siblings. When he was seven years old, Schultz's father broke his ankle and lost his job driving a diaper truck. That was all it took to throw the family into crisis. His father, after his ankle healed, began cycling through a series of lower-paying jobs. "My dad never found his way," Schultz told me. "I saw his self-esteem get battered. I felt like there was so much more he could have accomplished."

Schultz's school was a wild, overcrowded place with asphalt playgrounds and kids playing football, basketball, softball, punch ball, slap ball, and any other game they could devise. If your team lost, it could take an hour to get another turn. So Schultz made sure his team always won, no matter the cost. He would come home with bloody scrapes on his elbows and knees, which his mother would gently rinse with a wet cloth. "You don't quit," she told him.

His competitiveness earned him a college football scholarship (he broke his jaw and never played a game), a communications degree, and eventually a job as a Xerox salesman in New York City. He'd wake up every morning, go to a new midtown office building, take the elevator to the top floor, and go door-to-door, politely inquiring if anyone was interested in toner or copy machines. Then he'd ride the elevator down one floor and start all over again.

By the early 1980s, Schultz was working for a plastics manufacturer when he noticed that a little-known retailer in Seattle was ordering an inordinate number of coffee drip cones. Schultz flew out and fell in love with the company. Two years later, when he heard that Starbucks, then just six stores, was for sale, he asked everyone he knew for money and bought it.

That was 1987. Within three years, there were eighty-four stores; within six years, more than a thousand. Today, there are seventeen thousand stores in more than fifty countries.

Why did Schultz turn out so different from all the other kids on that playground? Some of his old classmates are today cops and firemen in Brooklyn. Others are in prison. Schultz is worth more than $1 billion. He's been heralded as one of the greatest CEOs of the twentieth century. Where did he find the determination—the willpower—to climb from a housing project to a private jet?

"I don't really know," he told me. "My mom always said, 'You're going to be the first person to go to college, you're going to be a professional, you're going to make us all proud.' She would ask these little questions, 'How are you going to study tonight? What are you going to do tomorrow? How do you know you're ready for your test?' It trained me to set goals.

"I've been really lucky," he said. "And I really, genuinely believe that if you tell people that they have what it takes to succeed, they'll prove you right."

Schultz's focus on employee training and customer service made Starbucks into one of the most successful companies in the world. For years, he was personally involved in almost every aspect of how the company was run. In 2000, exhausted, he handed over day-to-day operations to other executives, at which point, Starbucks began to stumble. Within a few years, customers were complaining about the quality of the drinks and customer service. Executives, focused on a frantic expansion, often ignored the complaints. Employees grew unhappy. Surveys indicated people were starting to equate Starbucks with tepid coffee and empty smiles.

So Schultz stepped back into the chief executive position in 2008. Among his priorities was restructuring the company's training program to renew its focus on a variety of issues, including bolstering employees'—or "partners," in Starbucks' lingo—willpower and self-confidence. "We had to start earning customer and partner trust again," Schultz told me.

At about the same time, a new wave of studies was appearing that looked at the science of willpower in a slightly different way. Researchers had noticed that some people, like Travis, were able to create willpower habits relatively easily. Others, however, struggled, no matter how much training and support they received. What was causing the difference?

Mark Muraven, who was by then a professor at the University of Albany, set up a new experiment. He put undergraduates in a room that contained a plate of warm, fresh cookies and asked them to ignore the treats. Half the participants were treated kindly. "We ask that you please don't eat the cookies. Is that okay?" a researcher said. She then discussed the purpose of the experiment, explaining that it was to measure their ability to resist temptations. She thanked them for contributing their time. "If you have any suggestions or thoughts about how we can improve this experiment, please let me know. We want you to help us make this experience as good as possible."

The other half of the participants weren't coddled the same way. They were simply given orders.

"You *must not* eat the cookies," the researcher told them. She didn't explain the experiment's goals, compliment them, or show any interest in their feedback. She told them to follow the instructions. "We'll start now," she said.

The students from both groups had to ignore the warm cookies for five minutes after the researcher left the room. None gave in to temptation.

Then the researcher returned. She asked each student to look at a computer monitor. It was programmed to flash numbers on the screen, one at a time, for five hundred milliseconds apiece. The participants were asked to hit the space bar every time they saw a "6" followed by a "4." This has become a standard way to measure willpower—paying attention to a boring sequence of flashing numbers requires a focus akin to working on an impossible puzzle.

Students who had been treated kindly did well on the computer test. Whenever a "6" flashed and a "4" followed, they pounced on the space bar. They were able to maintain their focus for the entire twelve minutes. Despite ignoring the cookies, they had willpower to spare.

Students who had been treated rudely, on the other hand, did terribly. They kept forgetting to hit the space bar. They said they were tired and couldn't focus. Their willpower muscle, researchers determined, had been fatigued by the brusque instructions.

When Muraven started exploring why students who had been treated kindly had more willpower he found that the key difference was the sense of control they had over their experience. "We've found this again and again," Muraven told me. "When people are asked to do something that takes self-control, if they think they are doing it for personal reasons—if they feel like it's a choice or something they enjoy because it helps someone else—it's much less taxing. If they feel like they have no autonomy, if they're just following orders, their willpower muscles get tired much faster. In both cases, people ignored the cookies. But when the students were treated like cogs, rather than people, it took a lot more willpower."

For companies and organizations, this insight has enormous implications. Simply giving employees a sense of agency—a feeling that they are in control, that they have genuine decision-making authority—can radically increase how much energy and focus they bring to their jobs. One 2010 study at a manufacturing plant in Ohio, for instance, scrutinized assembly-line workers who were empowered to make small decisions about their schedules and work environment. They designed their own uniforms and had authority over shifts. Nothing else changed. All the manufacturing processes and pay scales stayed the same. Within two months, productivity at the plant increased by 20 percent. Workers were taking shorter breaks. They were making fewer mistakes. Giving employees a sense of control improved how much self-discipline they brought to their jobs.

The same lessons hold true at Starbucks. Today, the company is focused on giving employees a greater sense of authority. They have asked workers to redesign how espresso machines and cash registers are laid out, to decide for themselves how customers should be greeted and where merchandise should be displayed. It's not unusual for a store manager to spend hours discussing with his employees where a blender should be located.

"We've started asking partners to use their intellect and creativity, rather than telling them 'take the coffee out of the box, put the cup here, follow this rule,'" said Kris Engskov, a vice president at Starbucks. "People want to be in control of their lives."

Turnover has gone down. Customer satisfaction is up. Since Schultz's return, Starbucks has boosted revenues by more than $1.2 billion per year.

V.

When Travis was sixteen, before he dropped out of school and started working for Starbucks, his mother told him a story. They were driving together, and Travis asked why he didn't have more siblings. His mother had always tried to be completely honest with her children, and so she told him that she had become pregnant two years before Travis was born but had gotten an abortion. They already had two children at that point, she explained, and were addicted to drugs. They didn't think they could support another baby. Then, a year later, she became pregnant with Travis. She thought about having another abortion, but it was too much to bear. It was easier to let nature take its course. Travis was born.

"She told me that she had made a lot of mistakes, but that having me was one of the best things that ever happened to her," Travis said. "When your parents are addicts, you grow up knowing you can't always trust them for everything you need. But I've been really lucky to find bosses who gave me what was missing. If my mom had been as lucky as me, I think things would have turned out different for her."

A few years after that conversation, Travis's father called to say that an infection had entered his mother's bloodstream through one of the places on her arm she used to shoot up. Travis immediately drove to the hospital in Lodi, but she was unconscious by the time he arrived. She died a half hour later, when they removed her life support.

A week later, Travis's father was in the hospital with pneumonia. His lung had collapsed. Travis drove to Lodi again, but it was 8:02 P.M.

when he got to the emergency room. A nurse brusquely told him he'd have to come back tomorrow; visiting hours were over.

Travis has thought a lot about that moment since then. He hadn't started working at Starbucks yet. He hadn't learned how to control his emotions. He didn't have the habits that, since then, he's spent years practicing. When he thinks about his life now, how far he is from a world where overdoses occur and stolen cars show up in driveways and a nurse seems like an insurmountable obstacle, he wonders how it's possible to travel such a long distance in such a short time.

"If he had died a year later, everything would have been different," Travis told me. By then, he would have known how to calmly plead with the nurse. He would have known to acknowledge her authority, and then ask politely for one small exception. He could have gotten inside the hospital. Instead, he gave up and walked away. "I said, 'All I want to do is talk to him once,' and she was like, 'He's not even awake, it's after visiting hours, come back tomorrow.' I didn't know what to say. I felt so small."

Travis's father died that night.

On the anniversary of his death, every year, Travis wakes up early, takes an extra-long shower, plans out his day in careful detail, and then drives to work. He always arrives on time.

6

THE POWER OF A CRISIS

How Leaders Create Habits Through
Accident and Design

I.

The patient was already unconscious when he was wheeled into the operating room at Rhode Island Hospital. His jaw was slack, his eyes closed, and the top of an intubation tube peeked above his lips. As a nurse hooked him up to a machine that would force air into his lungs during surgery, one of his arms slipped off the gurney, the skin mottled with liver spots.

The man was eighty-six years old and, three days earlier, had fallen at home. Afterward, he had trouble staying awake and answering questions, and so eventually his wife called an ambulance. In the emergency room, a doctor asked him what happened, but the man kept nodding off in the middle of his sentences. A scan of his head revealed why: The fall had slammed his brain against his skull, causing what's known as a subdural hematoma. Blood was pooling within the left portion of his cranium, pushing against the delicate folds of tissue inside his skull. The fluid had been building for almost seventy-two hours, and those parts of the brain that controlled his breathing and heart were beginning to falter. Unless the blood was drained, the man would die.

At the time, Rhode Island Hospital was one of the nation's leading medical institutions, the main teaching hospital for Brown University and the only Level I trauma center in southeastern New England. Inside the tall brick and glass building, physicians had pioneered cutting-edge medical techniques, including the use of ultrasound waves to destroy tumors inside a patient's body. In 2002, the National Coalition on Health Care rated the hospital's intensive care unit as one of the finest in the country.

But by the time the elderly patient arrived, Rhode Island Hospital also had another reputation: a place riven by internal tensions. There were deep, simmering enmities between nurses and physicians. In 2000, the nurses' union had voted to strike after complaining that they were being forced to work dangerously long hours. More than three hundred of them stood outside the hospital with signs reading "Stop Slavery" and "They can't take away our pride."

"This place can be awful," one nurse recalled telling a reporter. "The doctors can make you feel like you're worthless, like you're disposable. Like you should be thankful to pick up after them."

Administrators eventually agreed to limit nurses' mandatory overtime, but tensions continued to rise. A few years later, a surgeon was preparing for a routine abdominal operation when a nurse called for a "time-out." Such pauses are standard procedure at most hospitals, a way for doctors and staff to make sure mistakes are avoided. The nursing staff at Rhode Island Hospital was insistent on time-outs, particularly since a surgeon had accidentally removed the tonsils of a girl who was supposed to have eye surgery. Time-outs were supposed to catch such errors before they occurred.

At the abdominal surgery, when the OR nurse asked the team to gather around the patient for a time-out and to discuss their plan, the doctor headed for the doors.

"Why don't you lead this?" the surgeon told the nurse. "I'm going to step outside for a call. Knock when you're ready."

"You're supposed to be here for this, Doctor," she replied.

"You can handle it," the surgeon said, as he walked toward the door.

"Doctor, I don't feel this is appropriate."

The doctor stopped and looked at her. "If I want your damn opinion, I'll ask for it," he said. "Don't ever question my authority again. If you can't do your job, get the hell out of my OR."

The nurse led the time-out, retrieved the doctor a few minutes later, and the procedure occurred without complication. She never contradicted a physician again, and never said anything when other safety policies were ignored.

"Some doctors were fine, and some were monsters," one nurse who worked at Rhode Island Hospital in the mid-2000s told me. "We called it the glass factory, because it felt like everything could crash down at any minute."

To deal with these tensions, the staff had developed informal rules—habits unique to the institution—that helped avert the most obvious conflicts. Nurses, for instance, always double-checked the orders of error-prone physicians and quietly made sure that correct doses were entered; they took extra time to write clearly on patients' charts, lest a hasty surgeon make the wrong cut. One nurse told me they developed a system of color codes to warn one another. "We put doctors' names in different colors on the whiteboards," she said. "Blue meant 'nice,' red meant 'jerk,' and black meant, 'whatever you do, don't contradict them or they'll take your head off.'"

Rhode Island Hospital was a place filled with a corrosive culture. Unlike at Alcoa, where carefully designed keystone habits surrounding worker safety had created larger and larger successes, inside Rhode Island Hospital, habits emerged on the fly among nurses seeking to offset physician arrogance. The hospital's routines weren't carefully thought out. Rather, they appeared by accident and spread through whispered warnings, until toxic patterns emerged. This can happen within any organization where habits aren't deliberately planned. Just as choosing the right keystone habits can create amazing change, the wrong ones can create disasters.

And when the habits within Rhode Island Hospital imploded, they caused terrible mistakes.

● ● ●

When the emergency room staff saw the brain scans of the eighty-six-year-old man with the subdural hematoma, they immediately paged the neurosurgeon on duty. He was in the middle of a routine spinal surgery, but when he got the page, he stepped away from the operating table and looked at images of the elderly man's head on a computer screen. The surgeon told his assistant—a nurse practitioner—to go to the emergency room and get the man's wife to sign a consent form approving surgery. He finished his spinal procedure. A half hour later, the elderly man was wheeled into the same operating theater.

Nurses were rushing around. The unconscious elderly man was placed on the table. A nurse picked up his consent form and medical chart.

"Doctor," the nurse said, looking at the patient's chart. "The consent form doesn't say where the hematoma is." The nurse leafed through the paperwork. There was no clear indication of which side of his head they were supposed to operate on.

Every hospital relies upon paperwork to guide surgeries. Before any cut is made, a patient or family member is supposed to sign a document approving each procedure and verifying the details. In a chaotic environment, where as many as a dozen doctors and nurses may handle a patient between the ER and the recovery suite, consent forms are the instructions that keep track of what is supposed to occur. No one is supposed to go into surgery without a signed and detailed consent.

"I saw the scans before," the surgeon said. "It was the right side of the head. If we don't do this quickly, he's gonna die."

"Maybe we should pull up the films again," the nurse said, moving toward a computer terminal. For security reasons, the hospital's computers locked after fifteen minutes of idling. It would take at least a minute for the nurse to log in and load the patient's brain scans onto the screen.

"We don't have time," the surgeon said. "They told me he's crashing. We've got to relieve the pressure."

"What if we find the family?" the nurse asked.

"If that's what you want, then call the fucking ER and find the family! In the meantime, I'm going to save his life." The surgeon grabbed the paperwork, scribbled "right" on the consent form, and initialed it.

"There," he said. "We have to operate immediately."

The nurse had worked at Rhode Island Hospital for a year. He understood the hospital's culture. This surgeon's name, the nurse knew, was often scribbled in black on the large whiteboard in the hallway, signaling that nurses should beware. The unwritten rules in this scenario were clear: The surgeon always wins.

The nurse put down the chart and stood aside as the doctor positioned the elderly man's head in a cradle that provided access to the right side of his skull and shaved and applied antiseptic to his head. The plan was to open the skull and suction out the blood pooling on top of his brain. The surgeon sliced away a flap of scalp, exposed the skull, and put a drill against the white bone. He began pushing until the bit broke through with a soft pop. He made two more holes and used a saw to cut out a triangular piece of the man's skull. Underneath was the dura, the translucent sheath surrounding the brain.

"Oh my God," someone said.

There was no hematoma. They were operating on the wrong side of the head.

"We need him turned!" the surgeon yelled.

The triangle of bone was replaced and reattached with small metal plates and screws, and the patient's scalp sewed up. His head was shifted to the other side and then, once again, shaved, cleansed, cut, and drilled until a triangle of skull could be removed. This time, the hematoma was immediately visible, a dark bulge that spilled like thick syrup when the dura was pierced. The surgeon vacuumed the blood and the pressure inside the old man's skull fell immediately. The surgery, which should have taken about an hour, had run almost twice as long.

Afterward, the patient was taken to the intensive care unit, but he never regained full consciousness. Two weeks later, he died.

A subsequent investigation said it was impossible to determine the precise cause of death, but the patient's family argued that the trauma of the medical error had overwhelmed his already fragile body, that the stress of removing two pieces of skull, the additional time in surgery, and the delay in evacuating the hematoma had pushed him over the edge. If not for the mistake, they claimed, he might still be alive. The hospital paid a settlement and the surgeon was barred from ever working at Rhode Island Hospital again.

Such an accident, some nurses later claimed, was inevitable. Rhode Island Hospital's institutional habits were so dysfunctional, it was only a matter of time until a grievous mistake occurred.* It's not just hospitals that breed dangerous patterns, of course. Destructive organizational habits can be found within hundreds of industries and at thousands of firms. And almost always, they are the

---

*The reporting in this chapter is based upon interviews with multiple people working at Rhode Island Hospital and involved in this incident some of whom provided different accounts of events. For details on responses from hospital representatives and the surgeon involved, please see the notes.

products of thoughtlessness, of leaders who avoid thinking about the culture and so let it develop without guidance. There are no organizations without institutional habits. There are only places where they are deliberately designed, and places where they are created without forethought, so they often grow from rivalries or fear.

But sometimes, even destructive habits can be transformed by leaders who know how to seize the right opportunities. Sometimes, in the heat of a crisis, the right habits emerge.

II.

When *An Evolutionary Theory of Economic Change* was first published in 1982, very few people outside of academia noticed. The book's bland cover and daunting first sentence—"In this volume we develop an evolutionary theory of the capabilities and behavior of business firms operating in a market environment, and construct and analyze a number of models consistent with that theory"— almost seemed designed to ward off readers. The authors, Yale professors Richard Nelson and Sidney Winter, were best known for a series of intensely analytic papers exploring Schumpeterian theory that even most PhD candidates didn't pretend to understand.

Within the world of business strategy and organizational theory, however, the book went off like a bombshell. It was soon hailed as one of the most important texts of the century. Economics professors started talking about it to their colleagues at business schools, who started talking to CEOs at conferences, and soon executives were quoting Nelson and Winter inside corporations as different as General Electric, Pfizer, and Starwood Hotels.

Nelson and Winter had spent more than a decade examining how companies work, trudging through swamps of data before arriving at their central conclusion: "Much of firm behavior," they wrote, is best "understood as a reflection of general habits and strategic orientations coming from the firm's past," rather than "the result of a detailed survey of the remote twigs of the decision tree."

Or, put in language that people use outside of theoretical economics, it may *seem* like most organizations make rational choices based on deliberate decision making, but that's not really how companies operate at all. Instead, firms are guided by long-held organizational habits, patterns that often emerge from thousands of employees' independent decisions. And these habits have more profound impacts than anyone previously understood.

For instance, it might seem like the chief executive of a clothing company made the decision last year to feature a red cardigan on the catalog's cover by carefully reviewing sales and marketing data. But, in fact, what really happened was that his vice president constantly trolls websites devoted to Japanese fashion trends (where red was hip last spring), and the firm's marketers routinely ask their friends which colors are "in," and the company's executives, back from their annual trip to the Paris runway shows, reported hearing that designers at rival firms were using new magenta pigments. All these small inputs, the result of uncoordinated patterns among executives gossiping about competitors and talking to their friends, got mixed into the company's more formal research and development routines until a consensus emerged: Red will be popular this year. No one made a solitary, deliberate decision. Rather, dozens of habits, processes, and behaviors converged until it seemed like red was the inevitable choice.

These organizational habits—or "routines," as Nelson and Winter called them—are enormously important, because without them, most companies would never get any work done. Routines provide the hundreds of unwritten rules that companies need to operate. They allow workers to experiment with new ideas without having to ask for permission at every step. They provide a kind of "organizational memory," so that managers don't have to reinvent the sales process every six months or panic each time a VP quits. Routines reduce uncertainty—a study of recovery efforts after earthquakes in Mexico and Los Angeles, for instance, found that the habits of relief workers (which they carried from disaster to disaster, and which included things such as establishing communication networks by hiring children to carry messages between neighborhoods) were absolutely critical, "because without them, policy formulation and implementation would be lost in a jungle of detail."

But among the most important benefits of routines is that they create truces between potentially warring groups or individuals within an organization.

Most economists are accustomed to treating companies as idyllic places where everyone is devoted to a common goal: making as much money as possible. Nelson and Winter pointed out that, in the real world, that's not how things work at all. Companies aren't big happy families where everyone plays together nicely. Rather, most workplaces are made up of fiefdoms where executives compete for power and credit, often in hidden skirmishes that make their own performances appear superior and their rivals' seem worse. Divisions compete for resources and sabotage each other to steal glory. Bosses pit their subordinates against one another so that no one can mount a coup.

Companies aren't families. They're battlefields in a civil war.

Yet despite this capacity for internecine warfare, most companies roll along relatively peacefully, year after year, because they have routines—habits—that create truces that allow everyone to set aside their rivalries long enough to get a day's work done.

Organizational habits offer a basic promise: If you follow the established patterns and abide by the truce, then rivalries won't destroy the company, the profits will roll in, and, eventually, everyone will get rich. A salesperson, for example, knows she can boost her bonus by giving favored customers hefty discounts in exchange for larger orders. But she also knows that if every salesperson gives away hefty discounts, the firm will go bankrupt and there won't be any bonuses to hand out. So a routine emerges: The salespeople all get together every January and agree to limit how many discounts they offer in order to protect the company's profits, and at the end of the year everyone gets a raise.

Or take a young executive gunning for vice president who, with one quiet phone call to a major customer, could kill a sale and sabotage a colleague's division, taking him out of the running for the promotion. The problem with sabotage is that even if it's good for you, it's usually bad for the firm. So at most companies, an unspoken compact emerges: It's okay to be ambitious, but if you play *too* rough, your peers will unite against you. On the other hand, if you focus on boosting your own department, rather than undermining your rival, you'll probably get taken care of over time.

ROUTINES CREATE TRUCES
THAT ALLOW WORK TO GET DONE

Routines and truces offer a type of rough organizational justice, and because of them, Nelson and Winter wrote, conflict within companies usually "follows largely predictable paths and stays within predictable bounds that are consistent with the ongoing routine. . . . The usual amount of work gets done, reprimands and compliments are delivered with the usual frequency. . . . Nobody is trying to steer the organizational ship into a sharp turn in the hope of throwing a rival overboard."

Most of the time, routines and truces work perfectly. Rivalries still exist, of course, but because of institutional habits, they're kept within bounds and the business thrives.

However, sometimes even a truce proves insufficient. Sometimes, as Rhode Island Hospital discovered, an unstable peace can be as destructive as any civil war.

● ● ●

Somewhere in your office, buried in a desk drawer, there's probably a handbook you received on your first day of work. It contains expense forms and rules about vacations, insurance options, and the company's organizational chart. It has brightly colored graphs describing different health care plans, a list of relevant phone numbers, and instructions on how to access your email or enroll in the 401(k).

Now, imagine what you would tell a new colleague who asked for advice about how to *succeed* at your firm. Your recommendations probably wouldn't contain anything you'd find in the company's handbook. Instead, the tips you would pass along—who is trustworthy; which secretaries have more clout than their bosses; how to manipulate the bureaucracy to get something done—are the habits you rely on every day to survive. If you could somehow diagram all your work habits—and the informal power structures, relationships, alliances, and conflicts they represent—and then overlay your diagram with diagrams prepared by your colleagues, it would create a map of your firm's secret hierarchy, a guide to who knows how to make things happen and who never seems to get ahead of the ball.

Nelson and Winter's routines—and the truces they make possible—are critical to every kind of business. One study from Utrecht University in the Netherlands, for instance, looked at routines within the world of high fashion. To survive, every fashion designer has to possess some basic skills: creativity and a flair for haute couture as a start. But that's not enough to succeed. What makes the difference between success or failure are a designer's routines—whether they have a system for getting Italian broadcloth before wholesalers' stocks sell out, a process for finding the best zipper and button seamstresses, a routine for shipping a dress to a store in ten days, rather than three weeks. Fashion is such a complicated business that, without the right processes, a new company will get bogged down with logistics, and once that happens, creativity ceases to matter.

And which new designers are most likely to have the right habits? The ones who have formed the right truces and found the right alliances. Truces are so important that new fashion labels usually succeed only if they are headed by people who left *other* fashion companies on good terms.

Some might think Nelson and Winter were writing a book on dry economic theory. But what they really produced was a guide to surviving in corporate America.

What's more, Nelson and Winter's theories also explain why things went so wrong at Rhode Island Hospital. The hospital had routines that created an uneasy peace between nurses and doctors—the whiteboards, for instance, and the warnings nurses whispered to one another were habits that established a baseline truce. These delicate pacts allowed the organization to function most of the time. But truces are only durable when they create real justice. If a truce is unbalanced—if the peace isn't real—then the routines often fail when they are needed most.

The critical issue at Rhode Island Hospital was that the nurses were the only ones giving up power to strike a truce. It was the nurses who double-checked patients' medications and made extra efforts to write clearly on charts; the nurses who absorbed abuse from stressed-out doctors; the nurses who helped separate kind physicians from the despots, so the rest of the staff knew who tolerated operating-room suggestions and who would explode if you opened your mouth. The doctors often didn't bother to learn the nurses' names. "The doctors were in charge, and we were underlings," one nurse told me. "We tucked our tails and survived."

The truces at Rhode Island Hospital were one-sided. So at those crucial moments—when, for instance, a surgeon was about to make a hasty incision and a nurse tried to intervene—the routines that could have prevented the accident crumbled, and the wrong side of an eighty-six-year-old man's head was opened up.

Some might suggest that the solution is more equitable truces. That if the hospital's leadership did a better job of allocating authority, a healthier balance of power might emerge and nurses and doctors would be forced into a mutual respect.

That's a good start. Unfortunately, it isn't enough. Creating successful organizations isn't just a matter of balancing authority. For an organization to work, leaders must cultivate habits that both create a real and balanced peace and, paradoxically, make it absolutely clear who's in charge.

III.

Philip Brickell, a forty-three-year-old employee of the London Underground, was inside the cavernous main hall of the King's Cross subway station on a November evening in 1987 when a commuter stopped him as he was collecting tickets and said there was a burning tissue at the bottom of a nearby escalator.

King's Cross was one of the largest, grandest, and most heavily trafficked of London's subway stops, a labyrinth of deep escalators, passageways, and tunnels, some of which were almost a century old. The station's escalators, in particular, were famous for their size and age. Some stretched as many as five stories into the ground and were built of wooden slats and rubber handrails, the same materials used to construct them decades earlier. More than a quarter million passengers passed through King's Cross every day on six different

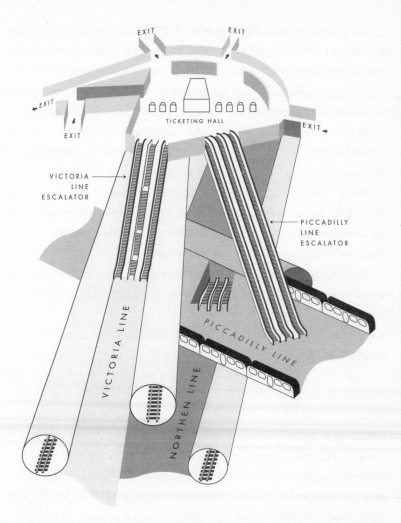

train lines. During evening rush hour, the station's ticketing hall was a sea of people hurrying beneath a ceiling repainted so many times that no one could recall its original hue.

The burning tissue, the passenger said, was at the bottom of one of the station's longest escalators, servicing the Piccadilly line. Brick-ell immediately left his position, rode the escalator down to the platform, found the smoldering wad of tissue, and, with a rolled-up magazine, beat out the fire. Then he returned to his post.

Brickell didn't investigate further. He didn't try to figure out why the tissue was burning or if it might have flown off of a larger fire somewhere else within the station. He didn't mention the incident to another employee or call the fire department. A separate department handled fire safety, and Brickell, in keeping with the strict divisions that ruled the Underground, knew better than to step on anyone's toes. Besides, even if he *had* investigated the possibility of a fire, he wouldn't have known what to do with any information he learned. The tightly prescribed chain of command at the Underground prohibited him from contacting another department without a superior's direct authorization. And the Underground's routines—handed down from employee to employee—told him that he should never, under any circumstances, refer out loud to *anything* inside a station as a "fire," lest commuters become panicked. It wasn't how things were done.

The Underground was governed by a sort of theoretical rule book that no one had ever seen or read—and that didn't, in fact, exist except in the unwritten rules that shaped every employee's life. For decades, the Underground had been run by the "Four Barons"—the chiefs of civil, signal, electrical, and mechanical engineering—and within each of their departments, there were bosses and subbosses who all jealously guarded their authority. The trains ran on time because all nineteen thousand Underground employees cooperated in a delicate system that passed passengers and trains among dozens—sometimes hundreds—of hands all day long. But that cooperation depended upon a balance of power between each of the four departments and all their lieutenants that, itself, relied upon thousands of habits that employees adhered to. These habits created a truce among the Four Barons and their deputies. And from that truce arose policies that told Brickell: Looking for fires isn't your job. Don't overstep your bounds.

"Even at the highest level, one director was unlikely to trespass on the territory of another," an investigator would later note. "Thus, the engineering director did not concern himself with whether the operating staff were properly trained in fire safety and evacuation procedures because he considered those matters to be the province of the Operations Directorate."

So Brickell didn't say anything about the burning tissue. In other circumstances, it might have been an unimportant detail. In this case, the tissue was a stray warning—a bit of fuel that had escaped from a larger, hidden blaze—that would show how perilous even perfectly balanced truces can become if they aren't designed just right.

Fifteen minutes after Brickell returned to his booth, another passenger noticed a wisp of smoke as he rode up the Piccadilly escalator; he mentioned it to an Underground employee. The King's Cross safety inspector, Christopher Hayes, was eventually roused to investigate. A third passenger, seeing smoke and a glow from underneath the escalator's stairs, hit an emergency stop button and began shouting at passengers to exit the escalator. A policeman saw a slight smoky haze inside the escalator's long tunnel, and, halfway down, flames beginning to dart above the steps.

Yet the safety inspector, Hayes, didn't call the London Fire Brigade. He hadn't seen any smoke himself, and another of the Underground's unwritten rules was that the fire department should never be contacted unless absolutely necessary. The policeman who had noticed the haze, however, figured he should contact headquarters. His radio didn't work underground, so he walked up a long staircase into the outdoors and called his superiors, who eventually passed word to the fire department. At 7:36 P.M.—twenty-two minutes after Brickell was alerted to the flaming tissue—the fire brigade received a call: "Small fire at King's Cross." Commuters were pushing past the policeman as he stood outside, speaking on his radio. They were rushing into the station, down into the tunnels, focused on getting home for dinner.

Within minutes, many of them would be dead.

●●●

At 7:36 P.M., an Underground worker roped off entry to the Picca-dilly escalator and another started diverting people to a different stairway. New trains were arriving every few minutes. The platforms where passengers exited subway cars were crowded. A bottleneck started building at the bottom of an open staircase.

Hayes, the safety inspector, went into a passageway that led to the Piccadilly escalator's machine room. In the dark, there was a set of controls for a sprinkler system specifically designed to fight fires on escalators. It had been installed years earlier, after a fire in an-other station had led to a series of dire reports about the risks of a sudden blaze. More than two dozen studies and reprimands had said that the Underground was unprepared for fires, and that staff needed to be trained in how to use sprinklers and fire extinguishers, which were positioned on every train platform. Two years earlier the deputy assistant chief of the London Fire Brigade had written to the operations director for railways, complaining about subway work-ers' safety habits.

"I am gravely concerned," the letter read. "I cannot urge too strongly that . . . clear instructions be given that on any suspicion of fire, the Fire Brigade be called without delay. This could save lives."

However, Hayes, the safety inspector, never saw that letter be-cause it was sent to a separate division from the one he worked within, and the Underground's policies were never rewritten to re-flect the warning. No one inside King's Cross understood how to use the escalator sprinkler system or was authorized to use the ex-tinguishers, because another department controlled them. Hayes completely forgot the sprinkler system existed. The truces ruling the Underground made sure everyone knew their place, but they left no room for learning about anything outside what you were as-signed to know. Hayes ran past the sprinkler controls without so much as a glance.

When he reached the machine room, he was nearly overcome by heat. The fire was already too big to fight. He ran back to the main hall. There was a line of people standing at the ticket machines and hundreds of people milling about the room, walking to platforms or leaving the station. Hayes found a policeman.

"We've got to stop the trains and get everyone out of here," he told him. "The fire is out of control. It's going everywhere."

At 7:42 P.M.—almost a half hour after the burning tissue—the first fireman arrived at King's Cross. As he entered the ticketing hall he saw dense black smoke starting to snake along the ceiling. The escalator's rubber handrails had begun to burn. As the acrid smell of burning rubber spread, commuters in the ticketing hall began to recognize that something was wrong. They moved toward the exits as firemen waded through the crowd, fighting against the tide.

Below, the fire was spreading. The entire escalator was now aflame, producing a superheated gas that rose to the top of the shaft enclosing the escalator, where it was trapped against the tunnel's ceiling, which was covered with about twenty layers of old paint. A few years earlier, the Underground's director of operations had suggested that all this paint might pose a fire hazard. Perhaps, he said, the old layers should be removed before a new one is applied?

Painting protocols were not in his purview, however. Paint responsibility resided with the maintenance department, whose chief politely thanked his colleague for the recommendation, and then noted that if he wanted to interfere with other departments, the favor would be swiftly returned.

The director of operations withdrew his recommendation.

As the superheated gases pooled along the ceiling of the escalator shaft, all those old layers of paint began absorbing the warmth. As each new train arrived, it pushed a fresh gust of oxygen into the station, feeding the fire like a bellows.

At 7:43 P.M., a train arrived and a salesman named Mark Silver exited. He knew immediately that something was wrong. The air was hazy, the platform packed with people. Smoke wafted around where he was standing, curling around the train cars as they sat on the tracks. He turned to reenter the train, but the doors had closed. He hammered on the windows, but there was an unofficial policy to avoid tardiness: Once the doors were sealed, they did not open again. Up and down the platform, Silver and other passengers screamed at the driver to open the doors. The signal light changed to green, and the train pulled away. One woman jumped on the tracks, running after the train as it moved into the tunnel. "Let me in!" she screamed.

Silver walked down the platform, to where a policeman was directing everyone away from the Piccadilly escalator and to another stairway. There were crowds of panicked people waiting to get upstairs. They could all smell the smoke, and everyone was packed together. It felt hot—either from the fire or the crush of people, Silver wasn't sure. He finally got to the bottom of an escalator that had been turned off. As he climbed toward the ticketing hall, he could feel his legs burning from heat coming through a fifteen-foot wall separating him from the Piccadilly shaft. "I looked up and saw the walls and ceiling sizzling," he later said.

At 7:45 P.M., an arriving train forced a large gust of air into the station. As the oxygen fed the fire, the blaze in the Piccadilly escalator roared. The superheated gases along the ceiling of the shaft, fueled by fire below and sizzling paint above, reached a combustion temperature, known as a "flashover point." At that moment, everything inside the shaft—the paint, the wooden escalator stairs, and any other available fuel—ignited in a fiery blast. The force of the sudden incineration acted the explosion of gunpowder at the base of a rifle barrel. It began pushing the fire upward through the long shaft, absorbing more heat and velocity as the blaze expanded until it shot out of the tunnel and into the ticketing hall in a wall of flames that set metal, tile, and flesh on fire. The temperature inside the hall shot up 150 degrees in half a second. A policeman riding one of the side escalators later told investigators that he saw "a jet of flame that shot up and then collected into a kind of ball." There were nearly fifty people inside the hall at the time.

Aboveground, on the street, a passerby felt heat explode from one of the subway's exits, saw a passenger stagger out, and ran to help. "I got hold of his right hand with my right hand but as our hands touched I could feel his was red hot and some of the skin came off in my hand," the rescuer said. A policeman who was entering the ticketing hall as the explosion occurred later told reporters, from a hospital bed, that "a fireball hit me in the face and knocked me off my feet. My hands caught fire. They were just melting."

He was one of the last people to exit the hall alive.

Shortly after the explosion, dozens of fire trucks arrived. But because the fire department's rules instructed them to connect their hoses to street-level hydrants, rather than those installed by the Underground inside the station, and because none of the subway employees had blueprints showing the station's layout—all the plans were in an office that was locked, and none of the ticketing agents or the station manager had keys—it took hours to extinguish the flames.

When the blaze was finally put out at 1:46 A.M.—six hours after the burning tissue was noticed—the toll stood at thirty-one dead and dozens injured.

"Why did they send me straight into the fire?" a twenty-year-old music teacher asked the next day from a hospital bed. "I could see them burning. I could hear them screaming. Why didn't someone take charge?"

● ● ●

To answer those questions, consider a few of the truces the London Underground relied upon to function:

Ticketing clerks were warned that their jurisdiction was strictly limited to selling tickets, so if they saw a burning tissue, they didn't warn anyone for fear of overstepping their bounds.

Station employees weren't trained how to use the sprinkler system or extinguishers, because that equipment was overseen by a different division.

The station's safety inspector never saw a letter from the London Fire Brigade warning about fire risks because it was sent to the operations director, and information like that wasn't shared across divisions.

Employees were instructed only to contact the fire brigade as a last resort, so as not to panic commuters unnecessarily.

The fire brigade insisted on using its own street-level hydrants, ignoring pipes in the ticketing hall that could have delivered water, because they had been ordered not to use equipment installed by other agencies.

In some ways, each of these informal rules, on its own, makes a certain amount of sense. For instance, the habits that kept ticketing clerks focused on selling tickets instead of doing anything else—including keeping an eye out for warning signs of fire—existed because, years earlier, the Underground had problems with understaffed kiosks. Clerks kept leaving their posts to pick up trash or point tourists toward their trains, and as a result, long lines would form. So clerks were ordered to stay in their booths, sell tickets, and not worry about anything else. It worked. Lines disappeared. If clerks saw something amiss outside their kiosks—beyond their scope of responsibility—they minded their own business.

And the fire brigade's habit of insisting on their own equipment? That was a result of an incident, a decade earlier, when a fire had raged in another station as firemen wasted precious minutes trying to hook up their hoses to unfamiliar pipes. Afterward, everyone decided it was best to stick with what they knew.

None of these routines, in other words, were arbitrary. Each was

designed for a reason. The Underground was so vast and complicated that it could operate smoothly only if truces smoothed over
potential obstacles. Unlike at Rhode Island Hospital, each truce created a genuine balance of power. No department had the upper
hand.

Yet thirty-one people died.

The London Underground's routines and truces all seemed logical until a fire erupted. At which point, an awful truth emerged: No
one person, department, or baron had ultimate responsibility for
passengers' safety.

Sometimes, one priority—or one department or one person or
one goal—*needs* to overshadow everything else, though it might be
unpopular or threaten the balance of power that keeps trains running on time. Sometimes, a truce can create dangers that outweigh
any peace.

There's a paradox in this observation, of course. How can an organization implement habits that balance authority and, at the same
time, choose a person or goal that rises above everyone else? How
do nurses and doctors share authority while still making it clear
who is in charge? How does a subway system avoid becoming
bogged down in turf battles while making sure safety is still a priority, even if that means lines of authority must be redrawn?

The answer lies in seizing the same advantage that Tony Dungy
encountered when he took over the woeful Bucs and Paul O'Neill
discovered when he became CEO of flailing Alcoa. It's the same opportunity Howard Schultz exploited when he returned to a flagging
Starbucks in 2007. All those leaders seized the possibilities created
by a crisis. During turmoil, organizational habits become malleable
enough to both assign responsibility and create a more equitable
balance of power. Crises are so valuable, in fact, that sometimes it's
worth stirring up a sense of looming catastrophe rather than letting
it die down.

IV.

Four months after the elderly man with the botched skull surgery died at Rhode Island Hospital, another surgeon at the hospital committed a similar error, operating on the wrong section of another patient's head. The state's health department reprimanded the facility and fined it $50,000. Eighteen months later, a surgeon operated on the wrong part of a child's mouth during a cleft palate surgery. Five months after that, a surgeon operated on a patient's wrong finger. Ten months after that, a drill bit was left inside a man's head. For these transgressions, the hospital was fined another $450,000.

Rhode Island Hospital is not the only medical institution where such accidents happen, of course, but they were unlucky enough to become the poster child for such mistakes. Local newspapers printed detailed stories of each incident. Television stations set up camp outside the hospital. The national media joined in, too. "The problem's not going away," a vice president of the national hospital accreditation organization told an Associated Press reporter. Rhode Island Hospital, the state's medical authorities declared to reporters, was a facility in chaos.

"It felt like working in a war zone," a nurse told me. "There were TV reporters ambushing doctors as they walked to their cars. One little boy asked me to make sure the doctor wouldn't accidentally cut off his arm during surgery. It felt like everything was out of control."

As critics and the media piled on, a sense of crisis emerged within the hospital. Some administrators started worrying that the facility would lose its accreditation. Others became defensive, attacking the television stations for singling them out. "I found a button that said 'Scapegoat' that I was going to wear to work," one doctor told me. "My wife said that was a bad idea."

Then an administrator, Dr. Mary Reich Cooper, who had become chief quality officer a few weeks before the eighty-six-year-old man's death, spoke up. In meetings with the hospital's administrators and staff, Cooper said that they were looking at the situation all wrong.

All this criticism wasn't a bad thing, she said. In fact, the hospital had been given an opportunity that few organizations ever received.

"I saw this as an opening," Dr. Cooper told me. "There's a long history of hospitals trying to attack these problems and failing. Sometimes people need a jolt, and all the bad publicity was a *serious* jolt. It gave us a chance to reexamine everything."

Rhode Island Hospital shut down all elective surgery units for an entire day—a huge expense—and put the entire staff through an intensive training program that emphasized teamwork and stressed the importance of empowering nurses and medical staff. The chief of neurosurgery resigned and a new leader was selected. The hospital invited the Center for Transforming Healthcare—a coalition of leading medical institutions—to help redesign its surgical safeguards. Administrators installed video cameras in operating rooms to make sure time-outs occurred and checklists were mandated for every surgery. A computerized system allowed any hospital employee to anonymously report problems that endangered patient health.

Some of those initiatives had been proposed at Rhode Island Hospital in previous years, but they had always been struck down. Doctors and nurses didn't want people recording their surgeries or other hospitals telling them how to do their jobs.

But once a sense of crisis gripped Rhode Island Hospital, everyone became more open to change.

Other hospitals have made similar shifts in the wake of mistakes and have brought down error rates that just years earlier had seemed immune to improvement. Like Rhode Island Hospital, these institutions have found that reform is usually possible only once a sense of crisis takes hold. For instance, one of Harvard University's teaching hospitals, Beth Israel Deaconess Medical Center, went through a spate of errors and internal battles in the late 1990s that spilled into newspaper articles and ugly shouting matches between nurses and administrators at public meetings. There was talk among some state officials of forcing the hospital to close departments until they could prove the mistakes would stop. Then the hospital, under attack, coalesced around solutions to change its culture. Part of the answer was "safety rounds," in which, every three months, a senior physician discussed a particular surgery or diagnosis and described, in painstaking detail, a mistake or near miss to an audience of hundreds of her or his peers.

"It's excruciating to admit a mistake publicly," said Dr. Donald Moorman, until recently Beth Israel Deaconess's associate surgeon in chief. "Twenty years ago, doctors wouldn't do it. But a real sense of panic has spread through hospitals now, and even the best surgeons are willing to talk about how close they came to a big error. The culture of medicine is changing."

●●●

Good leaders seize crises to remake organizational habits. NASA administrators, for instance, tried for years to improve the agency's safety habits, but those efforts were unsuccessful until the space shuttle *Challenger* exploded in 1986. In the wake of that tragedy, the organization was able to overhaul how it enforced quality standards. Airline pilots, too, spent years trying to convince plane manufacturers and air traffic controllers to redesign how cockpits were laid out and traffic controllers communicated. Then, a runway error on the Spanish island of Tenerife in 1977 killed 583 people and, within five years, cockpit design, runway procedures, and air traffic controller communication routines were overhauled.

In fact, crises are such valuable opportunities that a wise leader often prolongs a sense of emergency on purpose. That's exactly what occurred after the King's Cross station fire. Five days after the blaze, the British secretary of state appointed a special investigator, Desmond Fennell, to study the incident. Fennell began by interviewing the Underground's leadership, and quickly discovered that everyone had known—for years—that fire safety was a serious problem, and yet nothing had changed. Some administrators had proposed new hierarchies that would have clarified responsibility for fire prevention. Others had proposed giving station managers more power so that they could bridge departmental divides. None of those reforms had been implemented.

When Fennell began suggesting changes of his own, he saw the same kinds of roadblocks—department chiefs refusing to take responsibility or undercutting him with whispered threats to their subordinates—start to emerge.

So he decided to turn his inquiry into a media circus.

He called for public hearings that lasted ninety-one days and revealed an organization that had ignored multiple warnings of risks. He implied to newspaper reporters that commuters were in grave danger whenever they rode the subway. He cross-examined dozens of witnesses who described an organization where turf battles mattered more than commuter safety. His final report, released almost a year after the fire, was a scathing, 250-page indictment of the Underground portraying an organization crippled by bureaucratic ineptitude. "Having set out as an Investigation into the events of one night," Fennell wrote, the report's "scope was necessarily enlarged into the examination of a system." He concluded with pages and pages of stinging criticisms and recommendations that, essentially, suggested much of the organization was either incompetent or corrupt.

The response was instantaneous and overwhelming. Commuters picketed the Underground's offices. The organization's leadership was fired. A slew of new laws were passed and the culture of the Underground was overhauled. Today, every station has a manager whose primary responsibility is passenger safety, and every employee has an obligation to communicate at the smallest hint of risk. All the trains still run on time. But the Underground's habits

and truces have adjusted just enough to make it clear who has ultimate responsibility for fire prevention, and everyone is empowered
to act, regardless of whose toes they might step on.

The same kinds of shifts are possible at any company where institutional habits—through thoughtlessness or neglect—have created toxic truces. A company with dysfunctional habits can't turn
around simply because a leader orders it. Rather, wise executives
seek out moments of crisis—or create the perception of crisis—and
cultivate the sense that *something must change*, until everyone is finally ready to overhaul the patterns they live with each day.

"You never want a serious crisis to go to waste," Rahm Emanuel
told a conference of chief executives in the wake of the 2008 global
financial meltdown, soon after he was appointed as President
Obama's chief of staff. "This crisis provides the opportunity for us
to do things that you could not do before." Soon afterward, the
Obama administration convinced a once-reluctant Congress to pass
the president's $787 billion stimulus plan. Congress also passed
Obama's health care reform law, reworked consumer protection
laws, and approved dozens of other statutes, from expanding children's health insurance to giving women new opportunities to sue
over wage discrimination. It was one of the biggest policy overhauls
since the Great Society and the New Deal, and it happened because,
in the aftermath of a financial catastrophe, lawmakers saw opportunity.

Something similar happened at Rhode Island Hospital in the
wake of the eighty-six-year-old man's death and the other surgical
errors. Since the hospital's new safety procedures were fully implemented in 2009, no wrong-site errors have occurred. The hospital
recently earned a Beacon Award, the most prestigious recognition
of critical care nursing, and honors from the American College of
Surgeons for the quality of cancer care.

More important, say the nurses and doctors who work there,
Rhode Island Hospital feels like a completely different place.

In 2010, a young nurse named Allison Ward walked into an operating room to assist on a routine surgery. She had started working in the OR a year earlier. She was the youngest and least experienced person in the room. Before the surgery began, the entire surgical team gathered over the unconscious patient for a time-out. The surgeon read from a checklist, posted on the wall, which detailed every step of the operation.

"Okay, final step," he said before he picked up his scalpel. "Does anyone have any concerns before we start?"

The doctor had performed hundreds of these surgeries. He had an office full of degrees and awards.

"Doctor," the twenty-seven-year-old Ward said, "I want to remind everyone that we have to pause before the first and second procedures. You didn't mention that, and I just want to make sure we remember."

It was the type of comment that, a few years ago, might have earned her a rebuke. Or ended her career.

"Thanks for adding that," the surgeon said. "I'll remember to mention it next time.

"Okay," he said, "let's start."

"I know this hospital has gone through some hard periods," Ward later told me. "But it's really cooperative now. Our training, all the role models—the whole culture of the hospital is focused on teamwork. I feel like I can say anything. It's an amazing place to work."

7

HOW TARGET KNOWS WHAT YOU WANT BEFORE YOU DO

When Companies Predict (and Manipulate) Habits

I.

Andrew Pole had just started working as a data expert for Target when a few colleagues from the marketing department stopped by his desk one day and asked the kind of question Pole had been born to answer:

"Can your computers figure out which customers are pregnant, even if they don't want us to know?"

Pole was a statistician. His entire life revolved around using data to understand people. He had grown up in a small North Dakota town, and while his friends were attending 4-H or building model rockets, Pole was playing with computers. After college, he got a graduate degree in statistics and then another in economics, and while most of his classmates in the econ program at the University of Missouri were headed to insurance companies or government bureaucracies, Pole was on a different track. He'd become obsessed with the ways economists were using pattern analysis to explain human behavior. Pole, in fact, had tried his hand at a few informal experiments himself. He once threw a party and polled everyone on their favorite jokes, and then attempted to create a mathematical model for the perfect one-liner. He had sought to calculate the exact amount of beer he needed to drink in order to work up the confidence to talk to women at parties, but not so much that he would make a fool of himself. (That particular study never seemed to come out right.)

But those experiments were child's play, he knew, to how corporate America was using data to scrutinize people's lives. Pole wanted in. So when he graduated and heard that Hallmark, the greeting card company, was looking to hire statisticians in Kansas City, he submitted an application and was soon spending his days scouring sales data to determine if pictures of pandas or elephants sold more birthday cards, and if "What Happens at Grandma's Stays at Grandma's" is funnier in red or blue ink. It was heaven.

Six years later, in 2002, when Pole learned that Target was looking for number crunchers, he made the jump. Target, he knew, was a whole other magnitude when it came to data collection. Every year, millions of shoppers walked into Target's 1,147 stores and handed over terabytes of information about themselves. Most had no idea they were doing it. They used their customer loyalty cards, redeemed coupons they had received in the mail, or used a credit card, unaware that Target could then link their purchases to an individualized demographic profile.

To a statistician, this data was a magic window for peering into customers' preferences. Target sold everything from groceries to clothing, electronics and lawn furniture, and by closely tracking people's buying habits, the company's analysts could predict what was occurring within their homes. Someone's buying new towels, sheets, silverware, pans, and frozen dinners? They probably just bought a new house—or are getting a divorce. A cart loaded up with bug spray, kids' underwear, a flashlight, lots of batteries, *Real Simple,* and a bottle of Chardonnay? Summer camp is around the corner and Mom can hardly wait.

Working at Target offered Pole a chance to study the most com-plicated of creatures—the American shopper—in its natural habi-tat. His job was to build mathematical models that could crawl through data and determine which households contained kids and which were dedicated bachelors; which shoppers loved the outdoors and who was more interested in ice cream and romance novels. Pole's mandate was to become a mathematical mind reader, deci-phering shoppers' habits in order to convince them to spend more.

Then, one afternoon, a few of Pole's colleagues from the market-ing department stopped by his desk. They were trying to figure out which of Target's customers were pregnant based on their buying patterns, they said. Pregnant women and new parents, after all, are the holy grail of retail. There is almost no more profitable, product-hungry, price-insensitive group in existence. It's not just diapers and wipes. People with infants are so tired that they'll buy everything they need—juice and toilet paper, socks and magazines—wherever they purchase their bottles and formula. What's more, if a new par-ent starts shopping at Target, they'll keep coming back for years.

Figuring out who was pregnant, in other words, could make Tar-get millions of dollars.

Pole was intrigued. What better challenge for a statistical fortune-teller than not only getting inside shoppers' minds, but their bed-rooms?

By the time the project was done, Pole would learn some impor-tant lessons about the dangers of preying on people's most intimate habits. He would learn, for example, that hiding what you know is sometimes as important as knowing it, and that not all women are enthusiastic about a computer program scrutinizing their reproduc-tive plans.

Not everyone, it turns out, thinks mathematical mind reading is cool.

"I guess outsiders could say this is a little bit like Big Brother," Pole told me. "That makes some people uncomfortable."

● ● ●

Once upon a time, a company like Target would never have hired a guy like Andrew Pole. As little as twenty years ago retailers didn't do this kind of intensely data-driven analysis. Instead, Target, as well as grocery stores, shopping malls, greeting card sellers, clothing retailers, and other firms, tried to peer inside consumers' heads the old-fashioned way: by hiring psychologists who peddled vaguely scientific tactics they claimed could make customers spend more.

Some of those methods are still in use today. If you walk into a Walmart, Home Depot, or your local shopping center and look closely, you'll see retailing tricks that have been around for decades, each designed to exploit your shopping subconscious.

Take, for instance, how you buy food.

Chances are, the first things you see upon entering your grocery store are fruits and vegetables arranged in attractive, bountiful piles. If you think about it, positioning produce at the front of a store doesn't make much sense, because fruits and vegetables bruise easily at the bottom of a shopping cart; logically, they should be situated by the registers, so they come at the end of a trip. But as marketers and psychologists figured out long ago, if we *start* our shopping sprees by loading up on healthy stuff, we're much more likely to buy Doritos, Oreos, and frozen pizza when we encounter them later on. The burst of subconscious virtuousness that comes from first buying butternut squash makes it easier to put a pint of ice cream in the cart later.

Or take the way most of us turn to the right after entering a store. (Did you know you turn right? It's almost certain you do. There are thousands of hours of videotapes showing shoppers turning right once they clear the front doors.) As a result of this tendency, retailers fill the right side of the store with the most profitable products they're hoping you'll buy right off the bat. Or consider cereal and soups: When they're shelved out of alphabetical order and seemingly at random, our instinct is to linger a bit longer and look at a wider selection. So you'll rarely find Raisin Bran next to Rice Chex. Instead, you'll have to search the shelves for the cereal you want, and maybe get tempted to grab an extra box of another brand.

The problem with these tactics, however, is that they treat each shopper exactly the same. They're fairly primitive, one-size-fits-all solutions for triggering buying habits.

In the past two decades, however, as the retail marketplace has become more and more competitive, chains such as Target began to understand they couldn't rely on the same old bag of tricks. The only way to increase profits was to figure out each individual shopper's habits and to market to people one by one, with personalized pitches designed to appeal to customers' unique buying preferences.

In part, this realization came from a growing awareness of how powerfully habits influence almost every shopping decision. A series of experiments convinced marketers that if they managed to understand a particular shopper's habits, they could get them to buy almost anything. One study tape-recorded consumers as they walked through grocery stores. Researchers wanted to know how people made buying decisions. In particular, they looked for shoppers who had come with shopping lists—who, theoretically, had decided ahead of time what they wanted to get.

What they discovered was that despite those lists, more than 50 percent of purchasing decisions occurred at the moment a customer saw a product on the shelf, because, despite shoppers' best intentions, their habits were stronger than their written intentions. "Let's see," one shopper muttered to himself as he walked through a store. "Here are the chips. I will skip them. Wait a minute. Oh! The Lay's potato chips are on sale!" He put a bag in his cart. Some shoppers bought the same brands, month after month, even if they admitted they didn't like the product very much ("I'm not crazy about Folgers, but it's what I buy, you know? What else is there?" one woman said as she stood in front of a shelf containing dozens of other coffee brands). Shoppers bought roughly the same amount of food each time they went shopping, even if they had pledged to cut back.

"Consumers sometimes act like creatures of habit, automatically repeating past behavior with little regard to current goals," two psychologists at the University of Southern California wrote in 2009.

The surprising aspect of these studies, however, was that even though everyone relied on habits to guide their purchases, each person's habits were different. The guy who liked potato chips bought a bag every time, but the Folgers woman never went down the potato chip aisle. There were people who bought milk whenever they shopped—even if they had plenty at home—and there were people who always purchased desserts when they said they were trying to lose weight. But the milk buyers and the dessert addicts didn't usually overlap.

The habits were unique to each person.

Target wanted to take advantage of those individual quirks. But when millions of people walk through your doors every day, how do you keep track of their preferences and shopping patterns?

You collect data. Enormous, almost inconceivably large amounts of data.

Starting a little over a decade ago, Target began building a vast data warehouse that assigned every shopper an identification code—known internally as the "Guest ID number"—that kept tabs on how each person shopped. When a customer used a Target-issued credit card, handed over a frequent-buyer tag at the register, redeemed a coupon that was mailed to their house, filled out a survey, mailed in a refund, phoned the customer help line, opened an email from Target, visited Target.com, or purchased anything online, the company's computers took note. A record of each purchase was linked to that shopper's Guest ID number along with information on everything else they'd ever bought.

Also linked to that Guest ID number was demographic information that Target collected or purchased from other firms, including the shopper's age, whether they were married and had kids, which part of town they lived in, how long it took them to drive to the store, an estimate of how much money they earned, if they'd moved recently, which websites they visited, the credit cards they carried in their wallet, and their home and mobile phone numbers. Target can purchase data that indicates a shopper's ethnicity, their job history, what magazines they read, if they have ever declared bankruptcy, the year they bought (or lost) their house, where they went to college or graduate school, and whether they prefer certain brands of coffee, toilet paper, cereal, or applesauce.

There are data peddlers such as InfiniGraph that "listen" to shoppers' online conversations on message boards and Internet forums, and track which products people mention favorably. A firm named Rapleaf sells information on shoppers' political leanings, reading habits, charitable giving, the number of cars they own, and whether they prefer religious news or deals on cigarettes. Other companies analyze photos that consumers post online, cataloging if they are obese or skinny, short or tall, hairy or bald, and what kinds of products they might want to buy as a result. (Target, in a statement, declined to indicate what demographic companies it does business with and what kinds of information it studies.)

"It used to be that companies only knew what their customers *wanted* them to know," said Tom Davenport, one of the leading researchers on how businesses use data and analytics. "That world is far behind us. You'd be shocked how much information is out there—and every company buys it, because it's the only way to survive."

If you use your Target credit card to purchase a box of Popsicles once a week, usually around 6:30 P.M. on a weekday, and megasized trash bags every July and October, Target's statisticians and computer programs will determine that you have kids at home, tend to stop for groceries on your way back from work, and have a lawn that needs mowing in the summer and trees that drop leaves in the fall.

It will look at your other shopping patterns and notice that you sometimes buy cereal, but never purchase milk—which means that you must be buying it somewhere else. So Target will mail you coupons for 2 percent milk, as well as for chocolate sprinkles, school supplies, lawn furniture, rakes, and—since it's likely you'll want to relax after a long day at work—beer. The company will guess what you habitually buy, and then try to convince you to get it at Target. The firm has the capacity to personalize the ads and coupons it sends to every customer, even though you'll probably never realize you've received a different flyer in the mail than your neighbors.

"With the Guest ID, we have your name, address, and tender, we know you've got a Target Visa, a debit card, and we can tie that to your store purchases," Pole told an audience of retail statisticians at a conference in 2010. The company can link about half of all in-store sales to a specific person, almost all online sales, and about a quarter of online browsing.

At that conference, Pole flashed a slide showing a sample of the data Target collects, a diagram that caused someone in the audience to whistle in wonder when it appeared on the screen:

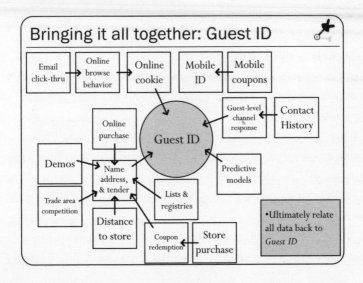

The problem with all this data, however, is that it's meaningless without statisticians to make sense of it. To a layperson, two shoppers who both buy orange juice look the same. It requires a special kind of mathematician to figure out that one of them is a thirty-four-year-old woman purchasing juice for her kids (and thus might appreciate a coupon for a Thomas the Tank Engine DVD) and the other is a twenty-eight-year-old bachelor who drinks juice after going for a run (and thus might respond to discounts on sneakers). Pole and the fifty other members of Target's Guest Data and Analytical Services department were the ones who found the habits hidden in the facts.

"We call it the 'guest portrait,'" Pole told me. "The more I know about someone, the better I can guess their buying patterns. I'm not going to guess everything about you every time, but I'll be right more often than I'm wrong."

By the time Pole joined Target in 2002, the analytics department had already built computer programs to identify households containing children and, come each November, send their parents catalogs of bicycles and scooters that would look perfect under the Christmas tree, as well as coupons for school supplies in September and advertisements for pool toys in June. The computers looked for shoppers buying bikinis in April, and sent them coupons for sunscreen in July and weight-loss books in December. If it wanted, Target could send each customer a coupon book filled with discounts for products they were fairly certain the shoppers were going to buy, because they had already purchased those exact items before.

Target isn't alone in its desire to predict consumers' habits. Almost every major retailer, including Amazon.com, Best Buy, Kroger supermarkets, 1-800-Flowers, Olive Garden, Anheuser-Busch, the U.S. Postal Service, Fidelity Investments, Hewlett-Packard, Bank of America, Capital One, and hundreds of others, have "predictive analytics" departments devoted to figuring out consumers' preferences. "But Target has always been one of the smartest at this," said Eric

Siegel, who runs a conference called Predictive Analytics World. "The data doesn't mean anything on its own. Target's good at figuring out the really clever questions."

It doesn't take a genius to know that someone buying cereal probably also needs milk. But there were other, much harder—and more profitable—questions to be answered.

Which is why, a few weeks after Pole was hired, his colleagues asked if it was possible to determine who was pregnant, even if that woman didn't want anyone to know.

● ● ●

In 1984, a visiting professor at UCLA named Alan Andreasen published a paper that set out to answer a basic question: Why do some people suddenly change their shopping routines?

Andreasen's team had spent the previous year conducting telephone surveys with consumers around Los Angeles, interrogating them about their recent shopping trips. Whenever someone answered the phone, the scientists would barrage them with questions about which brands of toothpaste and soap they had purchased and if their preferences had shifted. All told, they interviewed almost three hundred people. Like other researchers, they found that most people bought the same brands of cereal and deodorant week after week. Habits reigned supreme.

Except when they didn't.

For instance, 10.5 percent of the people Andreasen surveyed had switched toothpaste brands in the previous six months. More than 15 percent had started buying a new kind of laundry detergent.

Andreasen wanted to know why these people had deviated from their usual patterns. What he discovered has become a pillar of modern marketing theory: People's buying habits are more likely to change when they go through a major life event. When someone gets married, for example, they're more likely to start buying a new type of coffee. When they move into a new house, they're more apt to purchase a different kind of cereal. When they get divorced, there's a higher chance they'll start buying different brands of beer. Consumers going through major life events often don't notice, or care, that their shopping patterns have shifted. However, retailers notice, and they care quite a bit.

"Changing residence, getting married or divorced, losing or changing a job, having someone enter or leave the household," Andreasen wrote, are life changes that make consumers more "vulnerable to intervention by marketers."

And what's the biggest life event for most people? What causes the greatest disruption and "vulnerability to marketing interventions"? Having a baby. There's almost no greater upheaval for most customers than the arrival of a child. As a result, new parents' habits are more flexible at that moment than at almost any other period in an adult's life.

So for companies, pregnant women are gold mines.

New parents buy lots of stuff—diapers and wipes, cribs and Onesies, blankets and bottles—that stores such as Target sell at a significant profit. One survey conducted in 2010 estimated that the average parent spends $6,800 on baby items before a child's first birthday.

But that's just the tip of the shopping iceberg. Those initial expenditures are peanuts compared with the profits a store can earn by taking advantage of a new parent's shifting shopping habits. If exhausted moms and sleep-deprived dads start purchasing baby formula and diapers at Target, they'll start buying their groceries, cleaning supplies, towels, underwear, and—well, the sky's the limit—from Target as well. Because it's easy. To a new parent, easy matters most of all.

"As soon as we get them buying diapers from us, they're going to start buying everything else, too," Pole told me. "If you're rushing through the store, looking for bottles, and you pass orange juice, you'll grab a carton. Oh, and there's that new DVD I want. Soon, you'll be buying cereal and paper towels from us, and keep coming back."

New parents are so valuable that major retailers will do almost anything to find them, including going inside maternity wards, even if their products have nothing to do with infants. One New York hospital, for instance, provides every new mother with a gift bag containing samples of hair gel, face wash, shaving cream, an energy bar, shampoo, and a soft-cotton T-shirt. Inside are coupons for an online photo service, hand soap, and a local gym. There are also samples of diapers and baby lotions, but they're lost among the nonbaby supplies. In 580 hospitals across the United States, new mothers get gifts from the Walt Disney Company, which in 2010 started a division specifically aimed at marketing to the parents of infants. Procter & Gamble, Fisher-Price, and other firms have similar giveaway programs. Disney estimates the North American new baby market is worth $36.3 billion a year.

But for companies such as Target, approaching new moms in the maternity ward is, in some senses, too late. By then, they're already on everyone else's radar screen. Target didn't want to compete with Disney and Procter & Gamble; they wanted to beat them. Target's goal was to start marketing to parents *before* the baby arrived—which is why Andrew Pole's colleagues approached him that day to ask about building a pregnancy-prediction algorithm. If they could identify expecting mothers as early as their second trimester, they could capture them before anyone else.

The only problem was that figuring out which customers are pregnant is harder than it seems. Target had a baby shower registry, and that helped identify some pregnant women—and what's more, all those soon-to-be mothers willingly handed over valuable information, like their due dates, that let the company know when to send them coupons for prenatal vitamins or diapers. But only a fraction of Target's pregnant customers used the registry.

Then there were other customers who executives *suspected* were pregnant because they purchased maternity clothing, nursery furniture, and boxes of diapers. Suspecting and knowing, however, are two different things. How do you know whether someone buying diapers is pregnant or buying a gift for a pregnant friend? What's more, timing matters. A coupon that's useful a month before the due date might get put in the trash a few weeks after the baby arrives.

Pole started working on the problem by scouring the information in Target's baby shower registry, which let him observe how the average woman's shopping habits changed as her due date approached. The registry was like a laboratory where he could test hunches. Each expectant mother handed over her name, her spouse's name, and her due date. Target's data warehouse could link that information to the family's Guest IDs. As a result, whenever one of these women purchased something in a store or online, Pole, using the due date the woman provided, could plot the trimester in which the purchase occurred. Before long, he was picking up patterns.

Expectant mothers, he discovered, shopped in fairly predictable ways. Take, for example, lotions. Lots of people buy lotion, but a Target data analyst noticed that women on the baby registry were buying unusually large quantities of unscented lotion around the beginning of their second trimester. Another analyst noted that sometime in the first twenty weeks, many pregnant women loaded up on vitamins, such as calcium, magnesium, and zinc. Lots of shoppers purchase soap and cotton balls every month, but when someone suddenly starts buying lots of scent-free soap and cotton balls, in addition to hand sanitizers and an astounding number of washcloths, all at once, a few months after buying lotions and magnesium and zinc, it signals they are getting close to their delivery date.

As Pole's computer program crawled through the data, he was

able to identify about twenty-five different products that, when analyzed together, allowed him to, in a sense, peer inside a woman's womb. Most important, he could guess what trimester she was in—and estimate her due date—so Target could send her coupons when she was on the brink of making new purchases. By the time Pole was done, his program could assign almost any regular shopper a "pregnancy prediction" score.

Jenny Ward, a twenty-three-year-old in Atlanta who bought cocoa butter lotion, a purse large enough to double as a diaper bag, zinc, magnesium, and a bright blue rug? There's an 87 percent chance that she's pregnant and that her delivery date is sometime in late August. Liz Alter in Brooklyn, a thirty-five-year-old who purchased five packs of washcloths, a bottle of "sensitive skin" laundry detergent, baggy jeans, vitamins containing DHA, and a slew of moisturizers? She's got a 96 percent chance of pregnancy, and she'll probably give birth in early May. Caitlin Pike, a thirty-nine-year-old in San Francisco who purchased a $250 stroller, but nothing else? She's probably buying for a friend's baby shower. Besides, her demographic data shows she got divorced two years ago.

Pole applied his program to every shopper in Target's database. When it was done, he had a list of hundreds of thousands of women who were likely to be pregnant that Target could inundate with advertisements for diapers, lotions, cribs, wipes, and maternity clothing at times when their shopping habits were particularly flexible. If a fraction of those women or their husbands started doing their shopping at Target, it would add millions to the company's bottom line.

Then, just as this advertising avalanche was about to begin, someone within the marketing department asked a question: How are women going to react when they figure out how much Target knows?

"If we send someone a catalog and say, 'Congratulations on your first child!' and they've never told us they're pregnant, that's going to make some people uncomfortable," Pole told me. "We are very conservative about compliance with all privacy laws. But even if you're following the law, you can do things where people get queasy."

There's good reason for such worries. About a year after Pole created his pregnancy prediction model, a man walked into a Minnesota Target and demanded to see the manager. He was clutching an advertisement. He was very angry.

"My daughter got this in the mail!" he said. "She's still in high school, and you're sending her coupons for baby clothes and cribs? Are you trying to *encourage* her to get pregnant?"

The manager didn't have any idea what the man was talking about. He looked at the mailer. Sure enough, it was addressed to the man's daughter and contained advertisements for maternity clothing, nursery furniture, and pictures of smiling infants gazing into their mothers' eyes.

The manager apologized profusely, and then called, a few days later, to apologize again.

The father was somewhat abashed.

"I had a talk with my daughter," he said. "It turns out there's been some activities in my house I haven't been completely aware of." He took a deep breath. "She's due in August. I owe you an apology."

Target is not the only firm to have raised concerns among consumers. Other companies have been attacked for using data in far less intrusive ways. In 2011, for instance, a New York resident sued McDonald's, CBS, Mazda, and Microsoft, alleging those companies' advertising agency monitored people's Internet usage to profile their buying habits. There are ongoing class action lawsuits in California against Target, Walmart, Victoria's Secret, and other retail chains for asking customers to give their zip codes when they use credit cards, and then using that information to ferret out their mailing addresses.

Using data to predict a woman's pregnancy, Pole and his colleagues knew, was a potential public relations disaster. So how could they get their advertisements into expectant mothers' hands without making it appear they were spying on them? How do you take advantage of someone's habits without letting them know you're studying every detail of their lives?*

II.

In the summer of 2003, a promotion executive at Arista Records named Steve Bartels began calling up radio DJs to tell them about a new song he was certain they would love. It was called "Hey Ya!" by the hip-hop group OutKast.

"Hey Ya!" was an upbeat fusion of funk, rock, and hip-hop with a dollop of Big Band swing, from one of the most popular bands on earth. It sounded like nothing else on the radio. "It made the hair on my arms stand up the first time I heard it," Bartels told me. "It

---

*The reporting in this chapter is based on interviews with more than a dozen current and former Target employees, many of them conducted on a not-for-attribution basis because sources feared dismissal from the company or other retribution. Target was provided with an opportunity to review and respond to the reporting in this chapter, and was asked to make executives involved in the Guest Analytics department available for on-the-record interviews. The company declined to do so and declined to respond to fact-checking questions except in two emails. The first said: "At Target, our mission is to make Target the preferred shopping destination for our guests by delivering outstanding value, continuous innovation and an exceptional guest experience by consistently fulfilling our 'Expect More. Pay Less.' brand promise. Because we are so intently focused on this mission, we have made considerable investments in understanding our guests' preferences. To assist in this effort, we've developed a number of research tools that allow us to gain insights into trends and preferences within different demographic segments of our guest population. We use data derived from these tools to inform our store layouts, product selection, promotions and coupons. This analysis allows Target to provide the most relevant shopping experience to our guests. For example, during an in-store transaction, our research tool can predict relevant offers for an individual guest based on their purchases, which can be delivered along with their receipt. Further, opt-in programs such as our baby registry help Target understand how guests' needs evolve over time, enabling us to provide new mothers with money-saving coupons. We believe these efforts directly benefit our guests by providing more of what they need and want at Target—and have benefited Target by building stronger guest loyalty, driving greater shopping frequency and delivering increased sales and profitability." A second email read: "Almost all of your statements contain inaccurate information and publishing them would be misleading to the public. We do not intend to address each statement point by point. Target takes its legal obligations seriously and is in compliance with all applicable federal and state laws, including those related to protected health information."

sounded like a hit, like the kind of song you'd be hearing at bar mitz-vahs and proms for years." Around the Arista offices, executives sang the chorus—"shake it like a Polaroid picture"—to one another in the hallways. *This song,* they all agreed, *is going to be huge.*

That certainty wasn't based solely on intuition. At the time, the record business was undergoing a transformation similar to the data-driven shifts occurring at Target and elsewhere. Just as retailers were using computer algorithms to forecast shoppers' habits, music and radio executives were using computer programs to forecast lis-teners' habits. A company named Polyphonic HMI—a collection of artificial intelligence experts and statisticians based in Spain—had created a program called Hit Song Science that analyzed the math-ematical characteristics of a tune and predicted its popularity. By comparing the tempo, pitch, melody, chord progression, and other factors of a particular song against the thousands of hits stored in Polyphonic HMI's database, Hit Song Science could deliver a score that forecasted if a tune was likely to succeed.

The program had predicted that Norah Jones's *Come Away with Me,* for instance, would be a hit after most of the industry had dis-missed the album. (It went on to sell ten million copies and win eight Grammys.) It had predicted that "Why Don't You and I" by Santana would be popular, despite DJs' doubts. (It reached number three on the *Billboard* Top 40 list.)

When executives at radio stations ran "Hey Ya!" through Hit Song Science, it did well. In fact, it did better than well: The score was among the highest anyone had ever seen.

"Hey Ya!," according to the algorithm, was going to be a monster hit.

On September 4, 2003, in the prominent slot of 7:15 P.M., the Top 40 station WIOQ in Philadelphia started playing "Hey Ya!" on the radio. It aired the song seven more times that week, and a total of thirty-seven times throughout the month.

At the time, a company named Arbitron was testing a new technology that made it possible to figure out how many people were listening to a particular radio station at a given moment, and how many switched channels during a specific song. WIOQ was one of the stations included in the test. The station's executives were certain "Hey Ya!" would keep listeners glued to their radios.

Then the data came back.

Listeners didn't just dislike "Hey Ya!" They hated it according to the data. They hated it so much that nearly a third of them changed the station within the first thirty seconds of the song. It wasn't only at WIOQ, either. Across the nation, at radio stations in Chicago, Los Angeles, Phoenix, and Seattle, whenever "Hey Ya!" came on, huge numbers of listeners would click off.

"I thought it was a great song the first time I heard it," said John Garabedian, the host of a syndicated Top 40 radio show heard by more than two million people each weekend. "But it didn't sound like other songs, and so some people went nuts when it came on. One guy told me it was the worst thing he had ever heard.

"People listen to Top 40 because they want to hear their favorite songs or songs that sound just like their favorite songs. When something different comes on, they're offended. They don't want anything unfamiliar."

Arista had spent a lot of money promoting "Hey Ya!" The music and radio industries needed it to be a success. Hit songs are worth a fortune—not only because people buy the song itself, but also because a hit can convince listeners to abandon video games and the Internet for radio. A hit can sell sports cars on television and clothing inside trendy stores. Hit songs are at the root of dozens of spending habits that advertisers, TV stations, bars, dance clubs—even technology firms such as Apple—rely on.

Now, one of the most highly anticipated songs—a tune that the algorithms had predicted would become the song of the year—was flailing. Radio executives were desperate to find something that would make "Hey Ya!" into a hit.

●●●

That question—how do you make a song into a hit?—has been puzzling the music industry ever since it began, but it's only in the past few decades that people have tried to arrive at scientific answers. One of the pioneers was a onetime station manager named Rich Meyer who, in 1985, with his wife, Nancy, started a company called Mediabase in the basement of their Chicago home. They would wake up every morning, pick up a package of tapes of stations that had been recorded the previous day in various cities, and count and analyze every song that had been played. Meyer would then publish a weekly newsletter tracking which tunes were rising or declining in popularity.

In his first few years, the newsletter had only about a hundred subscribers, and Meyer and his wife struggled to keep the company afloat. However, as more and more stations began using Meyer's insights to increase their audiences—and, in particular, studying the formulas he devised to explain listening trends—his newsletter, the data sold by Mediabase, and then similar services provided by a growing industry of data-focused consultants, overhauled how radio stations were run.

One of the puzzles Meyer most loved was figuring out why, during some songs, listeners never seemed to change the radio dial. Among DJs, these songs are known as "sticky." Meyer had tracked hundreds of sticky songs over the years, trying to divine the principles that made them popular. His office was filled with charts and graphs plotting the characteristics of various sticky songs. Meyer was always looking for new ways to measure stickiness, and about the time "Hey Ya!" was released, he started experimenting with data from the tests that Arbitron was conducting to see if it provided any fresh insights.

Some of the stickiest songs at the time were sticky for obvious reasons—"Crazy in Love" by Beyoncé and "Señorita" by Justin Timberlake, for instance, had just been released and were already hugely popular, but those were great songs by established stars, so the stickiness made sense. Other songs, though, were sticky for reasons no one could really understand. For instance, when stations played "Breathe" by Blu Cantrell during the summer of 2003, almost no one changed the dial. The song is an eminently forgettable, beat-driven tune that DJs found so bland that most of them only played it reluctantly, they told music publications. But for some reason, whenever it came on the radio, people listened, even if, as pollsters later discovered, those same listeners said they didn't like the song very much. Or consider "Here Without You" by 3 Doors Down, or almost any song by the group Maroon 5. Those bands are so featureless that critics and listeners created a new music category—"bath rock"—to describe their tepid sounds. Yet whenever they came on the radio, almost no one changed the station.

Then there were songs that listeners said they actively *disliked*, but were sticky nonetheless. Take Christina Aguilera or Celine Dion. In survey after survey, male listeners said they hated Celine Dion and couldn't stand her songs. But whenever a Dion tune came on the radio, men stayed tuned in. Within the Los Angeles market, stations that regularly played Dion at the end of each hour—when the number of listeners was measured—could reliably boost their audience by as much as 3 percent, a huge figure in the radio world. Male listeners may have *thought* they disliked Dion, but when her songs played, they stayed glued.

One night, Meyer sat down and started listening to a bunch of sticky songs in a row, one right after the other, over and over again. As he did, he started to notice a similarity among them. It wasn't that the songs sounded alike. Some of them were ballads, others were pop tunes. However, they all seemed similar in that each sounded exactly like what Meyer expected to hear from that particular genre. They sounded *familiar*—like everything else on the radio—but a little more polished, a bit closer to the golden mean of the perfect song.

"Sometimes stations will do research by calling listeners on the phone, and play a snippet of a song, and listeners will say, 'I've heard that a million times. I'm totally tired of it,'" Meyer told me. "But when it comes on the radio, your subconscious says, 'I know this song! I've heard it a million times! I can sing along!' Sticky songs are what you *expect* to hear on the radio. Your brain secretly wants that song, because it's so familiar to everything else you've already heard and liked. It just sounds right."

There is evidence that a preference for things that sound "familiar" is a product of our neurology. Scientists have examined people's brains as they listen to music, and have tracked which neural regions are involved in comprehending aural stimuli. Listening to music activates numerous areas of the brain, including the auditory cortex, the thalamus, and the superior parietal cortex. These same areas are also associated with pattern recognition and helping the brain decide which inputs to pay attention to and which to ignore. The areas that process music, in other words, are designed to seek out patterns and look for familiarity. This makes sense. Music, after all, is complicated. The numerous tones, pitches, overlapping melodies, and competing sounds inside almost any song—or anyone speaking on a busy street, for that matter—are so overwhelming that, without our brain's ability to focus on some sounds and ignore others, everything would seem like a cacophony of noise.

Our brains crave familiarity in music because familiarity is how we manage to hear without becoming distracted by all the sound. Just as the scientists at MIT discovered that behavioral habits prevent us from becoming overwhelmed by the endless decisions we would otherwise have to make each day, listening habits exist because, without them, it would be impossible to determine if we should concentrate on our child's voice, the coach's whistle, or the noise from a busy street during a Saturday soccer game. Listening habits allow us to unconsciously separate important noises from those that can be ignored.

That's why songs that sound "familiar"—even if you've never heard them before—are sticky. Our brains are designed to prefer auditory patterns that seem similar to what we've already heard. When Celine Dion releases a new song—and it sounds like every other song she's sung, as well as most of the other songs on the radio—our brains unconsciously crave its recognizability and the song becomes sticky. You might never attend a Celine Dion concert, but you'll listen to her songs on the radio, because that's what you *expect* to hear as you drive to work. Those songs correspond perfectly to your habits.

This insight helped explain why "Hey Ya!" was failing on the radio, despite the fact that Hit Song Science and music executives were sure it would be a hit. The problem wasn't that "Hey Ya!" was bad. The problem was that "Hey Ya!" *wasn't familiar*. Radio listeners didn't want to make a conscious decision each time they were presented with a new song. Instead, their brains wanted to follow a habit. Much of the time, we don't actually choose if we like or dislike a song. It would take too much mental effort. Instead, we react to the cues ("This sounds like all the other songs I've ever liked") and rewards ("It's fun to hum along!") and without thinking, we either start singing, or reach over and change the station.

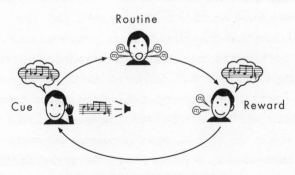

Routine

Cue

Reward

THE FAMILIARITY LOOP

In a sense, Arista and radio DJs faced a variation of the problem Andrew Pole was confronting at Target. Listeners are happy to sit through a song they might say they dislike, as long as it seems like something they've heard before. Pregnant women are happy to use coupons they receive in the mail, unless those coupons make it obvious that Target is spying into their wombs, which is unfamiliar and kind of creepy. Getting a coupon that makes it clear Target knows you're pregnant is at odds from what a customer expects. It's like telling a forty-two-year-old investment banker that he sang along to Celine Dion. It just feels wrong.

So how do DJs convince listeners to stick with songs such as "Hey Ya!" long enough for them to become familiar? How does Target convince pregnant women to use diaper coupons without creeping them out?

By dressing something new in old clothes, and making the unfamiliar seem familiar.

III.

In the early 1940s, the U.S. government began shipping much of the nation's domestic meat supply to Europe and the Pacific theater to support troops fighting in World War II. Back home, the availability of steaks and pork chops began to dwindle. By the time the United States entered the war in late 1941, New York restaurants were using horse meat for hamburgers and a black market for poultry had emerged. Federal officials became worried that a lengthy war effort would leave the nation starved of protein. This "problem will loom larger and larger in the United States as the war goes on," former president Herbert Hoover wrote to Americans in a government pamphlet in 1943. "Our farms are short of labor to care for livestock; and on top of it all we must furnish supplies to the British and Russians. Meats and fats are just as much munitions in this war as are tanks and aeroplanes."

Concerned, the Department of Defense approached dozens of the nation's leading sociologists, psychologists, and anthropologists—including Margaret Mead and Kurt Lewin, who would go on to become celebrity academics—and gave them an assignment: Figure out how to convince Americans to eat organ meats. Get housewives to serve their husbands and children the protein-rich livers, hearts, kidneys, brains, stomachs, and intestines that were left behind after the rib eyes and roast beef went overseas.

At the time, organ meat wasn't popular in America. A middle-class woman in 1940 would sooner starve than despoil her table with tongue or tripe. So when the scientists recruited into the Committee on Food Habits met for the first time in 1941, they set themselves a goal of systematically identifying the cultural barriers that discouraged Americans from eating organ meat. In all, more than two hundred studies were eventually published, and at their core, they all contained a similar finding: To change people's diets, the exotic must be made familiar. And to do that, you must camouflage it in everyday garb.

To convince Americans to eat livers and kidneys, housewives had to know how to make the foods look, taste, and smell as similar as possible to what their families *expected* to see on the dinner table, the scientists concluded. For instance, when the Subsistence Division of the Quartermaster Corps—the people in charge of feeding soldiers—started serving fresh cabbage to troops in 1943, it was rejected. So mess halls chopped and boiled the cabbage until it looked like every other vegetable on a soldier's tray—and the troops ate it without complaint. "Soldiers were more likely to eat food, whether familiar or unfamiliar, when it was prepared similar to their prior experiences and served in a familiar fashion," a present-day researcher evaluating those studies wrote.

The secret to changing the American diet, the Committee on Food Habits concluded, was familiarity. Soon, housewives were receiving mailers from the government telling them "every husband will cheer for steak and kidney pie." Butchers started handing out recipes that explained how to slip liver into meatloaf.

A few years after World War II ended, the Committee on Food Habits was dissolved. By then, however, organ meats had been fully integrated into the American diet. One study indicated that offal consumption rose by 33 percent during the war. By 1955, it was up 50 percent. Kidney had become a staple at dinner. Liver was for special occasions. America's dining patterns had shifted to such a degree that organ meats had become emblems of comfort.

Since then, the U.S. government has launched dozens of other efforts to improve our diets. For example, there was the "Five a Day" campaign, intended to encourage people to eat five fruits or vegetables, the USDA's food pyramid, and a push for low-fat cheeses and milks. None of them adhered to the committee's findings. None tried to camouflage their recommendations in existing habits, and as a result, all of the campaigns failed. To date, the only government program ever to cause a lasting change in the American diet was the organ meat push of the 1940s.

However, radio stations and massive companies—including Target—are a bit savvier.

● ● ●

To make "Hey Ya!" a hit, DJs soon realized, they needed to make the song feel familiar. And to do that, something special was required.

The problem was that computer programs such as Hit Song Science were pretty good at predicting people's habits. But sometimes, those algorithms found habits that hadn't actually emerged yet, and when companies market to habits we haven't adopted or, even worse, are unwilling to admit to ourselves—like our secret affection for sappy ballads—firms risk going out of business. If a grocery store boasts "We have a huge selection of sugary cereals and ice cream!"

shoppers stay away. If a butcher says "Here's a piece of intestine for your dinner table," a 1940s housewife serves tuna casserole instead. When a radio station boasts "Celine Dion every half hour!" no one tunes in. So instead, supermarket owners tout their apples and tomatoes (while making sure you pass the M&M's and Häagen-Dazs on the way to the register), butchers in the 1940s call liver "the new steak," and DJs quietly slip in the theme song from *Titanic*.

"Hey Ya!" needed to become part of an established listening habit to become a hit. And to become part of a habit, it had to be slightly camouflaged at first, the same way housewives camouflaged kidney by slipping it into meatloaf. So at WIOQ in Philadelphia—as well as at other stations around the nation—DJs started making sure that whenever "Hey Ya!" was played, it was sandwiched between songs that were already popular. "It's textbook playlist theory now," said Tom Webster, a radio consultant. "Play a new song between two consensus popular hits."

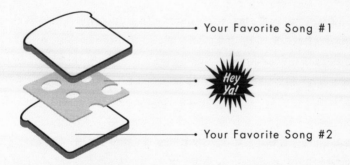

Your Favorite Song #1

Hey Ya!

Your Favorite Song #2

DJs, however, didn't air "Hey Ya!" alongside just any kind of hit. They sandwiched it between the types of songs that Rich Meyer had discovered were uniquely sticky, from artists like Blu Cantrell, 3 Doors Down, Maroon 5, and Christina Aguilera. (Some stations, in fact, were so eager they used the same song twice.)

Consider, for instance, the WIOQ playlist for September 19, 2003:

11:43	"Here Without You" by 3 Doors Down
11:54	"Breathe" by Blu Cantrell
11:58	"Hey Ya!" by OutKast
12:01	"Breathe" by Blu Cantrell

Or the playlist for October 16:

9:41	"Harder to Breathe" by Maroon 5
9:45	"Hey Ya!" by OutKast
9:49	"Can't Hold Us Down" by Christina Aguilera
10:00	"Frontin'" by Pharrell

November 12:

9:58	"Here Without You" by 3 Doors Down
10:01	"Hey Ya!" by OutKast
10:05	"Like I Love You" by Justin Timberlake
10:09	"Baby Boy" by Beyoncé

"Managing a playlist is all about risk mitigation," said Webster. "Stations have to take risks on new songs, otherwise people stop listening. But what listeners really want are songs they already like. So you have to make new songs seem familiar as fast as possible."

When WIOQ first started playing "Hey Ya!" in early September—before the sandwiching started—26.6 percent of listeners changed the station whenever it came on. By October, after playing it alongside sticky hits, that "tune-out factor" dropped to 13.7 percent. By December, it was 5.7 percent. Other major radio stations around the nation used the same sandwiching technique, and the tune-out rate followed the same pattern.

And as listeners heard "Hey Ya!" again and again, it became familiar. Once the song had become popular, WIOQ was playing "Hey Ya!" as many as fifteen times a day. People's listening habits had shifted to expect—crave, even—"Hey Ya!" A "Hey Ya!" habit emerged. The song went on to win a Grammy, sell more than 5.5

million albums, and earn radio stations millions of dollars. "This album cemented OutKast in the pantheon of superstars," Bartels, the promotion executive, told me. "This is what introduced them to audiences outside of hip-hop. It's so fulfilling now when a new artist plays me their single and says, *This is going to be the next 'Hey Ya!'*"

● ● ●

After Andrew Pole built his pregnancy-prediction machine, after he identified hundreds of thousands of female shoppers who were probably pregnant, after someone pointed out that some—in fact, most—of those women might be a little upset if they received an advertisement making it obvious Target knew their reproductive status, everyone decided to take a step back and consider their options.

The marketing department thought it might be wise to conduct a few small experiments before rolling out a national campaign. They had the ability to send specially designed mailers to small groups of customers, so they randomly chose women from Pole's pregnancy list and started testing combinations of advertisements to see how shoppers reacted.

"We have the capacity to send every customer an ad booklet, specifically designed for them, that says, 'Here's everything you bought last week, and a coupon for it,'" one Target executive with firsthand knowledge of Pole's pregnancy predictor told me. "We do that for grocery products all the time.

"With the pregnancy products, though, we learned that some women react badly. Then we started mixing in all these ads for things we knew pregnant women would never buy, so the baby ads looked random. We'd put an ad for a lawnmower next to diapers. We'd put a coupon for wineglasses next to infant clothes. That way, it looked like all the products were chosen by chance.

"And we found out that as long as a pregnant woman thinks she hasn't been spied on, she'll use the coupons. She just assumes that everyone else on her block got the same mailer for diapers and cribs. As long as we don't spook her, it works."

The answer to Target and Pole's question—how do you advertise to a pregnant woman without revealing that you know she's pregnant?—was essentially the same one that DJs used to hook listeners on "Hey Ya!" Target started sandwiching the diaper coupons between nonpregnancy products that made the advertisements seem anonymous, familiar, comfortable. They camouflaged what they knew.

Soon, Target's "Mom and Baby" sales exploded. The company doesn't break out sales figures for specific divisions, but between 2002—when Pole was hired—and 2009, Target's revenues grew from $44 billion to $65 billion. In 2005, the company's president, Gregg Steinhafel, boasted to a room full of investors about the company's "heightened focus on items and categories that appeal to specific guest segments such as mom and baby.

"As our database tools grow increasingly sophisticated, Target Mail has come into its own as a useful tool for promoting value and convenience to specific guest segments such as new moms or teens," he said. "For example, Target Baby is able to track life stages from prenatal care to car seats and strollers. In 2004, the Target Baby Direct Mail Program drove sizable increases in trips and sales."

Whether selling a new song, a new food, or a new crib, the lesson is the same: If you dress a new something in old habits, it's easier for the public to accept it.

IV.

The usefulness of this lesson isn't limited to large corporations, government agencies, or radio companies hoping to manipulate our tastes. These same insights can be used to change how we live.

In 2000, for instance, two statisticians were hired by the YMCA—one of the nation's largest nonprofit organizations—to use the powers of data-driven fortune-telling to make the world a healthier place. The YMCA has more than 2,600 branches in the United States, most of them gyms and community centers. About a decade ago, the organization's leaders began worrying about how to stay competitive. They asked a social scientist and a mathematician— Bill Lazarus and Dean Abbott—for help.

The two men gathered data from more than 150,000 YMCA member satisfaction surveys that had been collected over the years and started looking for patterns. At that point, the accepted wisdom among YMCA executives was that people wanted fancy exercise equipment and sparkling, modern facilities. The YMCA had spent millions of dollars building weight rooms and yoga studios. When the surveys were analyzed, however, it turned out that while a facility's attractiveness and the availability of workout machines might have caused people to join in the first place, what got them to stay was something else.

Retention, the data said, was driven by emotional factors, such as whether employees knew members' names or said hello when they walked in. People, it turns out, often go to the gym looking for a human connection, not a treadmill. If a member made a friend at the YMCA, they were much more likely to show up for workout sessions. In other words, people who join the YMCA have certain social habits. If the YMCA satisfied them, members were happy. So if the YMCA wanted to encourage people to exercise, it needed to take advantage of patterns that already existed, and teach employees to remember visitors' names. It's a variation of the lesson learned by Target and radio DJs: to sell a new habit—in this case exercise—wrap it in something that people already know and like, such as the instinct to go places where it's easy to make friends.

"We're cracking the code on how to keep people at the gym," Lazarus told me. "People want to visit places that satisfy their social needs. Getting people to exercise in groups makes it more likely they'll stick with a workout. You can change the health of the nation this way."

Someday soon, say predictive analytics experts, it will be possible for companies to know our tastes and predict our habits better than we know ourselves. However, knowing that someone might prefer a certain brand of peanut butter isn't enough to get them to act on that preference. To market a new habit—be it groceries or aerobics—you must understand how to make the novel seem familiar.

The last time I spoke to Andrew Pole, I mentioned that my wife was seven months pregnant with our second child. Pole himself has children, and so we talked a bit about kids. My wife and I shop at Target on occasion, I said, and about a year earlier we had given the company our address, so we could start getting coupons in the mail. Recently, as my wife's pregnancy had progressed, I'd been noticing a subtle upswing in the number of advertisements for diapers, lotions, and baby clothes arriving at our house.

I was planning on using some of those coupons that very weekend, I told him. I was also thinking of buying a crib, and some drapes for the nursery, and maybe some Bob the Builder toys for my toddler. It was really helpful that Target was sending me exactly the right coupons for what I needed to buy.

"Just wait till the baby comes," Pole said. "We'll be sending you coupons for things you want before you even know you want them."

PART
THREE

The Habits of Societies

8

SADDLEBACK CHURCH AND THE MONTGOMERY BUS BOYCOTT

How Movements Happen

I.

The 6 P.M. Cleveland Avenue bus pulled to the curb and the petite forty-two-year-old African American woman in rimless glasses and a conservative brown jacket climbed on board, reached into her purse, and dropped a ten-cent fare into the till.

It was Thursday, December 1, 1955, in Montgomery, Alabama, and she had just finished a long day at Montgomery Fair, the department store where she worked as a seamstress. The bus was crowded and, by law, the first four rows were reserved for white passengers. The area where blacks were allowed to sit, in the back, was already full and so the woman—Rosa Parks—sat in a center row, right behind the white section, where either race could claim a seat.

As the bus continued on its route, more people boarded. Soon, all the rows were filled and some—including a white passenger—were standing in the aisle, holding on to an overhead bar. The bus driver, James F. Blake, seeing the white man on his feet, shouted at the black passengers in Parks's area to give up their seats, but no one moved. It was noisy. They might not have heard. Blake pulled over to a bus stop in front of the Empire Theater on Montgomery Street and walked back.

"Y'all better make it light on yourselves and let me have those seats," he said. Three of the black passengers got up and moved to the rear, but Parks stayed put. She wasn't *in* the white section, she told the driver, and besides, there was only one white rider standing.

"If you don't stand up," Blake said, "I'm going to call the police and have you arrested."

"You may do that," Parks said.

The driver left and found two policemen.

"Why don't you stand up?" one of them asked Parks after they boarded.

"Why do you push us around?" she said.

"I don't know," the officer answered. "But the law is the law and you're under arrest."

At that moment, though no one on that bus knew it, the civil rights movement pivoted. That small refusal was the first in a series of actions that shifted the battle over race relations from a struggle fought by activists in courts and legislatures into a contest that would draw its strength from entire communities and mass protests. Over the next year, Montgomery's black population would rise up and boycott the city's buses, ending their strike only once the law segregating races on public transportation was stricken from the books. The boycott would financially cripple the bus line, draw tens of thousands of protesters to rallies, introduce the country to a charismatic young leader named Martin Luther King, Jr., and spark a movement that would spread to Little Rock, Greensboro, Raleigh, Birmingham, and, eventually, to Congress. Parks would become a hero, a recipient of the Presidential Medal of Freedom, and a shining example of how a single act of defiance can change the world.

But that isn't the whole story. Rosa Parks and the Montgomery bus boycott became the epicenter of the civil rights campaign not only because of an individual act of defiance, but also because of social patterns. Parks's experiences offer a lesson in the power of social habits—the behaviors that occur, unthinkingly, across dozens or hundreds or thousands of people which are often hard to see as they emerge, but which contain a power that can change the world. Social habits are what fill streets with protesters who may not know one another, who might be marching for different reasons, but who are all moving in the same direction. Social habits are why some initiatives become world-changing movements, while others fail to ignite. And the reason why social habits have such influence is because at the root of many movements—be they large-scale revolutions or simple fluctuations in the churches people attend—is a three-part process that historians and sociologists say shows up again and again:

A movement starts because of the social habits of friendship and the strong ties between close acquaintances.

It grows because of the habits of a community, and the weak ties that hold neighborhoods and clans together.

And it endures because a movement's leaders give participants new habits that create a fresh sense of identity and a feeling of ownership.

Usually, only when all three parts of this process are fulfilled can a movement become self-propelling and reach a critical mass. There are other recipes for successful social change and hundreds of details that differ between eras and struggles. But understanding how social habits work helps explain why Montgomery and Rosa Parks became the catalyst for a civil rights crusade.

It wasn't inevitable that Parks's act of rebellion that winter day would result in anything other than her arrest. Then habits intervened, and something amazing occurred.

● ● ●

Rosa Parks wasn't the first black passenger jailed for breaking Montgomery's bus segregation laws. She wasn't even the first that year. In 1946, Geneva Johnson had been arrested for talking back to a Montgomery bus driver over seating. In 1949, Viola White, Katie Wingfield, and two black children were arrested for sitting in the white section and refusing to move. That same year, two black teenagers visiting from New Jersey—where buses were integrated—were arrested and jailed after breaking the law by sitting next to a white man and a boy. In 1952, a Montgomery policeman shot and killed a black man when he argued with a bus driver. In 1955, just months before Parks was taken to jail, Claudette Colvin and Mary Louise Smith were arrested in separate incidents for refusing to give their seats to white passengers.

None of those arrests resulted in boycotts or protests, however. "There weren't many real activists in Montgomery at the time," Taylor Branch, the Pulitzer Prize–winning civil rights historian, told me. "People didn't mount protests or marches. Activism was something that happened in courts. It wasn't something average people did."

When a young Martin Luther King, Jr., arrived in Montgomery in 1954, for instance, a year before Parks's arrest, he found a majority of the city's blacks accepted segregation "without apparent protest. Not only did they seem resigned to segregation per se; they also accepted the abuses and indignities which came with it."

So why, when Parks was arrested, did things change?

One explanation is that the political climate was shifting. The previous year, the U.S. Supreme Court had handed down *Brown v. Board of Education*, ruling that segregation was illegal within public schools; six months before Parks's arrest, the Court had issued what came to be known as *Brown II*—a decision ordering that school integration must proceed with "all deliberate speed." There was a powerful sense across the nation that change was in the air.

But that isn't sufficient to explain why Montgomery became

ground zero for the civil rights struggle. Claudette Colvin and Mary Louise Smith had been arrested in the wake of *Brown v. Board,* and yet they didn't spark a protest. *Brown,* for many Montgomery residents, was an abstraction from a far-off courthouse, and it was unclear how—or if—its impact would be felt locally. Montgomery wasn't Atlanta or Austin or other cities where progress seemed possible. "Montgomery was a pretty nasty place," Branch said. "Racism was set in its ways there."

When Parks was arrested, however, it sparked something unusual within the city. Rosa Parks, unlike other people who had been jailed for violating the bus segregation law, was deeply respected and embedded within her community. So when she was arrested, it triggered a series of social habits—the habits of friendship—that ignited an initial protest. Parks's membership in dozens of social networks across Montgomery allowed her friends to muster a response before the community's normal apathy could take hold.

Montgomery's civil life, at the time, was dominated by hundreds of small groups that created the city's social fabric. The city's *Directory of Civil and Social Organizations* was almost as thick as its phone book. Every adult, it seemed—particularly every black adult—belonged to some kind of club, church, social group, community center, or neighborhood organization, and often more than one. And within these social networks, Rosa Parks was particularly well known and liked. "Rosa Parks was one of those rare people of whom everyone agreed that she gave more than she got," Branch wrote in his history of the civil rights movement, *Parting the Waters.* "Her character represented one of the isolated high blips on the graph of human nature, offsetting a dozen or so sociopaths." Parks's many friendships and affiliations cut across the city's racial and economic lines. She was the secretary of the local NAACP chapter, attended the Methodist church, and helped oversee a youth organization at the Lutheran church near her home. She spent some weekends volunteering at a shelter, others with a botanical club, and on Wednesday nights often joined a group of women who knit blankets for a local hospital. She volunteered dressmaking services to poor families and provided last-minute gown alterations for wealthy white debutantes. She was so deeply enmeshed in the community, in fact, that her husband complained that she ate more often at potlucks than at home.

In general, sociologists say, most of us have friends who are like us. We might have a few close acquaintances who are richer, a few who are poorer, and a few of different races—but, on the whole, our deepest relationships tend to be with people who look like us, earn about the same amount of money, and come from similar backgrounds.

Parks's friends, in contrast, spanned Montgomery's social and economic hierarchies. She had what sociologists call "strong ties"—firsthand relationships—with dozens of groups throughout Montgomery that didn't usually come into contact with one another. "This was absolutely key," Branch said. "Rosa Parks transcended the social stratifications of the black community and Montgomery as a whole. She was friends with field hands and college professors."

And the power of those friendships became apparent as soon as Parks landed in jail.

● ● ●

Rosa Parks called her parents' home from the police station. She was panicked, and her mother—who had no idea what to do—started going through a mental Rolodex of Parks's friends, trying to think of someone who might be able to help. She called the wife of E. D. Nixon, the former head of the Montgomery NAACP, who in turn called her husband and told him that Parks needed to be bailed out of jail. He immediately agreed to help, and called a prominent white lawyer named Clifford Durr who knew Parks because she had hemmed dresses for his three daughters.

Nixon and Durr went to the jailhouse, posted bail for Parks, and took her home. They'd been looking for the perfect case to challenge Montgomery's bus segregation laws, and sensing an opportunity, they asked Parks if she would be willing to let them fight her arrest in court. Parks's husband was opposed to the idea. "The white folks will kill you, Rosa," he told her.

But Parks had spent years working with Nixon at the NAACP. She had been in Durr's house and had helped his daughters prepare for cotillions. Her friends were now asking her for a favor.

"If you think it will mean something to Montgomery and do some good," she told them, "I'll be happy to go along with it."

That night—just a few hours after the arrest—news of Parks's jailing began to filter through the black community. Jo Ann Robinson, the president of a powerful group of schoolteachers involved in politics and a friend of Parks's from numerous organizations, heard about it. So did many of the schoolteachers in Robinson's group, and many of the parents of their students. Close to midnight, Robinson called an impromptu meeting and suggested that everyone boycott the city's buses on Monday, four days hence, when Parks was to appear in court.

Afterward, Robinson snuck into her office's mimeograph room and made copies of a flyer.

"Another Negro woman has been arrested and thrown into jail because she refused to get up out of her seat on the bus for a white person to sit down," it read. "This woman's case will come up on Monday. We are, therefore, asking every Negro to stay off the buses Monday in protest of the arrest and trial."

Early the next morning, Robinson gave stacks of the flyers to schoolteachers and asked them to distribute it to parents and co-workers. Within twenty-four hours of Parks's arrest, word of her jailing and the boycott had spread to some of the city's most influential communities—the local NAACP, a large political group, a number of black schoolteachers, and the parents of their students. Many of the people who received a flyer knew Rosa Parks personally—they had sat next to her in church or at a volunteer meeting and considered her a friend. There's a natural instinct embedded in friendship, a sympathy that makes us willing to fight for someone we like when they are treated unjustly. Studies show that people have no problem ignoring strangers' injuries, but when a friend is insulted, our sense of outrage is enough to overcome the inertia that usually makes protests hard to organize. When Parks's friends learned about her arrest and the boycott, the social habits of friendship—the natural inclination to help someone we respect—kicked in.

The first mass movement of the modern civil rights era could have been sparked by any number of earlier arrests. But it began with Rosa Parks because she had a large, diverse, and connected set of friends—who, when she was arrested, reacted as friends naturally respond, by following the social habits of friendship and agreeing to show their support.

Still, many expected the protest would be nothing more than a one-day event. Small protests pop up every day around the world, and almost all of them quickly fizzle out. No one has enough friends to change the world.

Which is why the second aspect of the social habits of movements is so important. The Montgomery bus boycott became a society-wide action because the sense of obligation that held the black community together was activated soon after Parks's friends started spreading the word. People who hardly knew Rosa Parks decided to participate because of a social peer pressure—an influence known as "the power of weak ties"—that made it difficult to avoid joining in.

II.

Imagine, for a moment, that you're an established midlevel executive at a prosperous company. You're successful and well liked. You've spent years building a reputation inside your firm and cultivating a network of friends that you can tap for clients, advice, and industry gossip. You belong to a church, a gym, and a country club, as well as the local chapter of your college alumni association. You're respected and often asked to join various committees. When people within your community hear of a business opportunity, they often pass it your way.

Now imagine you get a phone call. It's a midlevel executive at another company looking for a new job. Will you help him by putting in a good word with your boss, he asks?

If the person on the telephone is a total stranger, it's an easy decision. Why risk your standing inside your firm helping someone you don't know?

If the person on the phone is a close friend, on the other hand, it's also an easy choice. Of course you'll help. That's what friends do.

However, what if the person on the phone isn't a good friend or a stranger, but something in between? What if you have friends in common, but don't know each other very well? Do you vouch for the caller when your boss asks if he's worth an interview? How much of your own reputation and energy, in other words, are you willing to expend to help a friend of a friend get a job?

In the late 1960s, a Harvard PhD student named Mark Granovetter set out to answer that question by studying how 282 men had found their current employment. He tracked how they had learned about open positions, whom they had called for referrals, the methods they used to land interviews, and most important, who had provided a helping hand. As expected, he found that when job hunters approached strangers for assistance, they were rejected. When they appealed to friends, help was provided.

More surprising, however, was how often job hunters also received help from casual acquaintances—friends of friends—people who were neither strangers nor close pals. Granovetter called those connections "weak ties," because they represented the links that connect people who have acquaintances in common, who share membership in social networks, but aren't directly connected by the strong ties of friendship themselves.

In fact, in landing a job, Granovetter discovered, weak-tie acquaintances were often *more* important than strong-tie friends because weak ties give us access to social networks where we don't otherwise belong. Many of the people Granovetter studied had learned about new job opportunities through weak ties, rather than from close friends, which makes sense because we talk to our closest friends all the time, or work alongside them or read the same blogs. By the time they have heard about a new opportunity, we probably know about it, as well. On the other hand, our weak-tie acquaintances—the people we bump into every six months—are the ones who tell us about jobs we would otherwise never hear about.

When sociologists have examined how opinions move through communities, how gossip spreads or political movements start, they've discovered a common pattern: Our weak-tie acquaintances are often as influential—if not more—than our close-tie friends. As Granovetter wrote, "Individuals with few weak ties will be deprived of information from distant parts of the social system and will be confined to the provincial news and views of their close friends. This deprivation will not only insulate them from the latest ideas and fashions but may put them in a disadvantaged position in the labor market, where advancement can depend . . . on knowing about appropriate job openings at just the right time.

"Furthermore, such individuals may be difficult to organize or integrate into political movements of any kind. . . . While members of one or two cliques may be efficiently recruited, the problem is that, without weak ties, any momentum generated in this way does not spread *beyond* the clique. As a result, most of the population will be untouched."

The power of weak ties helps explain how a protest can expand from a group of friends into a broad social movement. Convincing thousands of people to pursue the same goal—especially when that pursuit entails real hardship, such as walking to work rather than taking the bus, or going to jail, or even skipping a morning cup of coffee because the company that sells it doesn't support organic farming—is hard. Most people don't care enough about the latest outrage to give up their bus ride or caffeine unless it's a close friend that has been insulted or jailed. So there is a tool that activists have long relied upon to compel protest, even when a group of people don't necessarily *want* to participate. It's a form of persuasion that has been remarkably effective over hundreds of years. It's the sense of obligation that neighborhoods or communities place upon themselves.

In other words, peer pressure.

Peer pressure—and the social habits that encourage people to conform to group expectations—is difficult to describe, because it often differs in form and expression from person to person. These social habits aren't so much one consistent pattern as dozens of individual habits that ultimately cause everyone to move in the same direction.

The habits of peer pressure, however, have something in common. They often spread through weak ties. And they gain their authority through communal expectations. If you ignore the social obligations of your neighborhood, if you shrug off the expected patterns of your community, you risk losing your social standing. You endanger your access to many of the social benefits that come from joining the country club, the alumni association, or the church in the first place.

In other words, if you don't give the caller looking for a job a helping hand, he might complain to his tennis partner, who might mention those grumblings to someone in the locker room who you were hoping to attract as a client, who is now less likely to return your call because you have a reputation for not being a team player. On a playground, peer pressure is dangerous. In adult life, it's how business gets done and communities self-organize.

Such peer pressure, on its own, isn't enough to sustain a movement. But when the strong ties of friendship and the weak ties of peer pressure merge, they create incredible momentum. That's when widespread social change can begin.

● ● ●

To see how the combination of strong and weak ties can propel a movement, fast forward to nine years *after* Rosa Parks's arrest, when hundreds of young people volunteered to expose themselves to deadly risks for the civil rights crusade.

In 1964, students from across the country—many of them whites from Harvard, Yale, and other northern universities—applied for something called the "Mississippi Summer Project." It was a ten-week program devoted to registering black voters in the South. The project came to be known as Freedom Summer, and many who applied were aware it would be dangerous. In the months before the program started, newspapers and magazines were filled with articles predicting violence (which proved tragically accurate when, just a week after it began, white vigilantes killed three volunteers outside Longdale, Mississippi). The threat of harm kept many students from participating in the Mississippi Summer Project, even after they applied. More than a thousand applicants were accepted into Freedom Summer, but when it came time to head south in June, more than three hundred of those invited to participate decided to stay home.

In the 1980s, a sociologist at the University of Arizona named Doug McAdam began wondering if it was possible to figure out why some people had participated in Freedom Summer and others withdrew. He started by reading 720 of the applications students had submitted decades earlier. Each was five pages long. Applicants were asked about their backgrounds, why they wanted to go to Mississippi, and their experiences with voter registration. They were told to provide a list of people organizers should contact if they were arrested. There were essays, references, and, for some, interviews. Applying was not a casual undertaking.

McAdam's initial hypothesis was that students who ended up going to Mississippi probably had different motivations from those who stayed home, which explained the divergence in participation. To test this idea, he divided applicants into two groups. The first pile were people who said they wanted to go to Mississippi for "self-interested" motives, such as to "test myself," to "be where the action is," or to "learn about the southern way of life." The second group were those with "other-oriented" motives, such as to "improve the lot of blacks," to "aid in the full realization of democracy," or to "demonstrate the power of nonviolence as a vehicle for social change."

The self-centered, McAdam hypothesized, would be more likely to stay home once they realized the risks of Freedom Summer. The other-oriented would be more likely to get on the bus.

The hypothesis was wrong.

The selfish and the selfless, according to the data, went South in equal numbers. Differences in motives did not explain "any significant distinctions between participants and withdrawals," McAdam wrote.

Next, McAdam compared applicants' opportunity costs. Maybe those who stayed home had husbands or girlfriends keeping them from going to Mississippi? Maybe they had gotten jobs, and couldn't swing a two-month unpaid break?

Wrong again.

"Being married or holding a full-time job actually enhanced the applicant's chances of going south," McAdam concluded.

He had one hypothesis left. Each applicant was asked to list their memberships in student and political organizations and at least ten people they wanted kept informed of their summer activities, so McAdam took these lists and used them to chart each applicant's social network. By comparing memberships in clubs, he was able to determine which applicants had friends who also applied for Freedom Summer.

Once he finished, he finally had an answer as to why some students went to Mississippi, and others stayed home: because of social habits—or more specifically, because of the power of strong and weak ties working in tandem. The students who participated in Freedom Summer were enmeshed in the types of communities where both their close friends *and* their casual acquaintances expected them to get on the bus. Those who withdrew were also enmeshed in communities, but of a different kind—the kind where the social pressures and habits didn't compel them to go to Mississippi.

"Imagine you're one of the students who applied," McAdam told me. "On the day you signed up for Freedom Summer, you filled out the application with five of your closest friends and you were all feeling really motivated.

"Now, it's six months later and departure day is almost here. All the magazines are predicting violence in Mississippi. You called your parents, and they told you to stay at home. It would be strange, at that point, if you weren't having second thoughts.

"Then, you're walking across campus and you see a bunch of people from your church group, and they say, 'We're coordinating rides—when should we pick you up?' These people aren't your closest friends, but you see them at club meetings and in the dorm, and they're important within your social community. They all know you've been accepted to Freedom Summer, and that you've said you want to go. Good luck pulling out at that point. You'd lose a huge amount of social standing. Even if you're having second thoughts, there's real consequences if you withdraw. You'll lose the respect of people whose opinions matter to you."

When McAdam looked at applicants with religious orientations—students who cited a "Christian duty to help those in need" as their motivation for applying, for instance, he found mixed levels of participation. However, among those applicants who mentioned a religious orientation *and* belonged to a religious organization, McAdam found that *every single one* made the trip to Mississippi. Once their communities knew they had been accepted into Freedom Summer, it was impossible for them to withdraw.

On the other hand, consider the social networks of applicants who were accepted into the program but didn't go to Mississippi. They, too, were involved in campus organizations. They, too, belonged to clubs and cared about their standing within those communities. But the organizations they belonged to—the newspaper and student government, academic groups and fraternities—had different expectations. Within those communities, someone could withdraw from Freedom Summer and suffer little or no decline in the prevailing social hierarchy.

When faced with the prospect of getting arrested (or worse) in Mississippi, most students probably had second thoughts. However, some were embedded in communities where social habits—the expectations of their friends and the peer pressure of their acquaintances—compelled participation, so regardless of their hesitations, they bought a bus ticket. Others—who also cared about civil rights—belonged to communities where the social habits pointed in a slightly different direction, so they thought to themselves, *Maybe I'll just stay home.*

● ● ●

On the morning after he bailed Rosa Parks out of jail, E. D. Nixon placed a call to the new minister of the Dexter Avenue Baptist Church, Martin Luther King, Jr. It was a little after 5 A.M., but Nixon didn't say hello or ask if he had awoken King's two-week-old daughter when the minister answered—he just launched into an account of Parks's arrest, how she had been hauled into jail for refusing to

give up her seat, and their plans to fight her case in court and boycott the city's buses on Monday. At the time, King was twenty-six
years old. He had been in Montgomery for only a year and was still
trying to figure out his role within the community. Nixon was asking for King's endorsement as well as permission to use his church
for a boycott meeting that night. King was wary of getting too deeply
involved. "Brother Nixon," he said, "let me think about it and you
call me back."

But Nixon didn't stop there. He reached out to one of King's closest friends—one of the strongest of King's strong ties—named
Ralph D. Abernathy, and asked him to help convince the young
minister to participate. A few hours later, Nixon called King again.

"I'll go along with it," King told him.

"I'm glad to hear you say so," Nixon said, "because I've talked to
eighteen other people and told them to meet in your church tonight.
It would have been kind of bad to be getting together there without
you." Soon, King was drafted into serving as president of the organization that had sprung up to coordinate the boycott.

On Sunday, three days after Parks's arrest, the city's black
ministers—after speaking to King and other members of the new
organization—explained to their congregations that every black
church in the city had agreed to a one-day protest. The message was
clear: It would be embarrassing for any parishioner to sit on the
sidelines. That same day, the town's newspaper, the *Advertiser*, contained an article about "a 'top secret' meeting of Montgomery Negroes who plan a boycott of city buses Monday." The reporter had
gotten copies of flyers that white women had taken from their
maids. The black parts of the city were "flooded with thousands of
copies" of the leaflets, the article explained, and it was anticipated
that every black citizen would participate. When the article was written, only Parks's friends, the ministers, and the boycott organizers
had publicly committed to the protest—but once the city's black

residents read the newspaper, they assumed, like white readers, that everyone else was already on board.

Many people sitting in the pews and reading the newspapers knew Rosa Parks personally and were willing to boycott because of their friendships with her. Others didn't know Parks, but they could sense the community was rallying behind her cause, and that if they were seen riding a bus on Monday, it would look bad. "If you work," read a flyer handed out in churches, "take a cab, or share a ride, or walk." Then everyone heard that the boycott's leaders had convinced—or strong-armed—all the black taxi drivers into agreeing to carry black passengers on Monday for ten cents a ride, the same as a bus fare. The community's weak ties were drawing everyone together. At that point, you were either with the boycott or against it.

On the Monday morning of the boycott, King woke before dawn and got his coffee. His wife, Coretta, sat at the front window and waited for the first bus to pass. She shouted when she saw the headlights of the South Jackson line, normally filled with maids on their way to work, roll by with no passengers. The next bus was empty as well. And the one that came after. King got into his car and started driving around, checking other routes. In an hour, he counted eight black passengers. One week earlier, he would have seen hundreds.

"I was jubilant," he later wrote. "A miracle had taken place. . . . Men were seen riding mules to work, and more than one horse-drawn buggy drove the streets of Montgomery. . . . Spectators had gathered at the bus stops to watch what was happening. At first, they stood quietly, but as the day progressed they began to cheer the empty buses and laugh and make jokes. Noisy youngsters could be heard singing out, 'No riders today.'"

That afternoon, in a courtroom on Church Street, Rosa Parks was found guilty of violating the state's segregation laws. More than five hundred blacks crowded the hallways and stood in front of the building, awaiting the verdict. The boycott and impromptu rally at the courthouse were the most significant black political activism in Montgomery's history, and it had all come together in five days. It had started among Parks's close friends, but it drew its power, King and other participants later said, because of a sense of obligation among the community—the social habits of weak ties. The community was pressured to stand together for fear that anyone who didn't participate wasn't someone you wanted to be friends with in the first place.

There are plenty of people who would have participated in the boycott without such encouragement. King and the cabbies and the congregations might have made the same choices without the influence of strong and weak ties. But tens of thousands of people from across the city would not have decided to stay off the buses without the encouragement of social habits. "The once dormant and quiescent Negro community was now fully awake," King later wrote.

Those social habits, however, weren't strong enough on their own to extend a one-day boycott into a yearlong movement. Within a few weeks, King would be openly worrying that people's resolve was weakening, that "the ability of the Negro community to continue the struggle" was in doubt.

Then those worries would evaporate. King, like thousands of other movement leaders, would shift the struggle's guidance from his hands onto the shoulders of his followers, in large part by handing them new habits. He would activate the third part of the movement formula, and the boycott would become a self-perpetuating force.

III.

In the summer of 1979, a young seminary student who was white, had been one year old when Rosa Parks was arrested, and was currently focused mostly on how he was going to support his growing family, posted a map on the wall of his Texas home and began drawing circles around major U.S. cities, from Seattle to Miami.

Rick Warren was a Baptist pastor with a pregnant wife and less than $2,000 in the bank. He wanted to start a new congregation among people who didn't already attend church, but he had no idea where it should be located. "I figured I would go somewhere all my seminary friends didn't want to go," he told me. He spent the summer in libraries studying census records, phone books, newspaper articles, and maps. His wife was in her ninth month, and so every few hours Warren would jog to a pay phone, call home to make sure she hadn't started labor yet, and then return to the stacks.

One afternoon, Warren stumbled upon a description of a place called Saddleback Valley in Orange County, California. The book Warren was reading said it was the fastest-growing region in the fastest-growing county in one of the fastest-growing states in America. There were a number of churches in the area, but none large enough to accommodate the quickly expanding population. Intrigued, Warren contacted religious leaders in Southern California who told him that many locals self-identified as Christian but didn't attend services. "In the dusty, dimly lit basement of that university library, I heard God speak to me: 'That's where I want you to plant a church!'" Warren later wrote. "From that moment on, our destination was a settled issue."

Warren's focus on building a congregation among the unchurched had begun five years earlier, when, as a missionary in Japan, he had discovered an old copy of a Christian magazine with an article headlined "Why Is This Man Dangerous?" It was about Donald McGavran, a controversial author focused on building churches in nations where most people hadn't accepted Christ. At the center of McGavran's philosophy was an admonition that missionaries should imitate the tactics of other successful movements—including the civil rights campaign—by appealing to people's social habits. "The steady goal must be the Christianization of the entire fabric which is the people, or large enough parts of it that the social life of the individual is not destroyed," McGavran had written in one of his books. Only the evangelist who helps people "to become followers of Christ *in their normal social relationship* has any chance of liberating multitudes."

That article—and, later, McGavran's books—were a revelation to Rick Warren. Here, finally, was someone applying a rational logic to a topic that was usually couched in the language of miracles. Here was someone who understood that religion had to be, for lack of a better word, marketed.

McGavran laid out a strategy that instructed church builders to speak to people in their "own languages," to create places of worship where congregants saw their friends, heard the kinds of music they already listened to, and experienced the Bible's lessons in digestible metaphors. Most important, McGavran said, ministers needed to convert *groups* of people, rather than individuals, so that a community's social habits would encourage religious participation, rather than pulling people away.

In December, after graduating from seminary and having the baby, Warren loaded his family and belongings into a U-Haul, drove to Orange County, and rented a small condo. His first prayer group attracted all of seven people and took place in his living room.

Today, thirty years later, Saddleback Church is one of the largest ministries in the world, with more than twenty thousand parishioners visiting its 120-acre campus—and eight satellite campuses—each week. One of Warren's books, *The Purpose-Driven Life,* has sold thirty million copies, making it among the biggest sellers in history. There are thousands of other churches modeled on his methods. Warren was chosen to perform the invocation at President Obama's inauguration, and is considered one of the most influential religious leaders on earth.

And at the core of his church's growth and his success is a fundamental belief in the power of social habits.

"We've thought long and hard about habitualizing faith, breaking it down into pieces," Warren told me. "If you try to scare people into following Christ's example, it's not going to work for too long. The only way you get people to take responsibility for their spiritual maturity is to teach them *habits* of faith.

"Once that happens, they become self-feeders. People follow Christ not because you've led them there, but because it's who they are."

● ● ●

When Warren first arrived in Saddleback Valley, he spent twelve weeks going door-to-door, introducing himself and asking strangers why they *didn't* go to church. Many of the answers were practical—it was boring, people said, the music was bad, the sermons didn't seem applicable to their lives, they needed child care, they hated dressing up, the pews were uncomfortable.

Warren's church would address each of those complaints. He told people to wear shorts and Hawaiian shirts, if they felt like it. An electric guitar was brought in. Warren's sermons, from the start, focused on practical topics, with titles such as "How to Handle Discouragement," "How to Feel Good About Yourself," "How to Raise Healthy Families," and "How to Survive Under Stress." His lessons were easy to understand, focused on real, daily problems, and could be applied as soon as parishioners left church.

It started to work. Warren rented school auditoriums for services and office buildings for prayer meetings. The congregation hit fifty members, then one hundred, then two hundred in less than a year. Warren was working eighteen hours a day, seven days a week, answering congregants' phone calls, leading classes, coming to their homes to offer marriage counseling, and, in his spare time, always looking for new venues to accommodate the church's growing size.

One Sunday in mid-December, Warren stood up to preach during the eleven o'clock service. He felt light-headed, dizzy. He gripped the podium and started to speak, but the words on the page were blurry. He began to fall, caught himself, and motioned to the assistant pastor—his only staff—to take the lectern.

"I'm sorry, folks," Warren told the audience. "I'm going to have to sit down."

For years, he had suffered from anxiety attacks and occasional bouts of melancholy that friends told him sounded like mild depressions. But it had never hit this bad before. The next day, Warren and his family began driving to Arizona, where his wife's family had a house. Slowly, he recuperated. Some days, he would sleep for twelve hours and then take a walk through the desert, praying, trying to understand why these panic attacks were threatening to undo everything he had worked so hard to build. Nearly a month passed as he stayed away from the church. His melancholy became a full-fledged depression, darker than anything he had experienced before. He wasn't certain if he would ever become healthy enough to return.

Warren, as befitting a pastor, is a man prone to epiphanies. They had occurred when he found the magazine article about McGavran, and in the library in Texas. Walking through the desert, another one struck.

"You focus on building people," the Lord told him. "And I will build the church."

Unlike some of his previous revelations, however, this one didn't suddenly make the path clear. Warren would continue to struggle with depression for months—and then during periods throughout his life. On that day, however, he made two decisions: He would go back to Saddleback, and he would figure out how to make running the church less work.

● ● ●

When Warren returned to Saddleback, he decided to expand a small experiment he had started a few months earlier that, he hoped, would make it easier to manage the church. He was never certain he would have enough classrooms to accommodate everyone who showed up for Bible study, so he had asked a few church members to host classes inside their homes. He worried that people might complain about going to someone's house, rather than a proper church classroom. But congregants loved it, they said. The small groups gave them a chance to meet their neighbors. So, after he returned from his leave, Warren assigned every Saddleback member to a small group that met every week. It was one of the most important decisions he ever made, because it transformed church participation from a decision into a habit that drew on already-existing social urges and patterns.

"Now, when people come to Saddleback and see the giant crowds on the weekends, they think that's our success," Warren told me. "But that's just the tip of the iceberg. Ninety-five percent of this church is what happens during the week inside those small groups.

"The congregation and the small groups are like a one-two punch. You have this big crowd to remind you why you're doing this in the first place, and a small group of close friends to help you focus on how to be faithful. Together, they're like glue. We have over five thousand small groups now. It's the only thing that makes a church this size manageable. Otherwise, I'd work myself to death, and 95 percent of the congregation would never receive the attention they came here looking for."

Without realizing it, Warren, in some ways, has replicated the structure that propelled the Montgomery bus boycott—though he has done it in reverse. That boycott started among people who knew Rosa Parks, and became a mass protest when the weak ties of the community compelled participation. At Saddleback Church, it works the other way around. People are attracted by a sense of community and the weak ties that a congregation offers. Then once inside, they're pushed into a small group of neighbors—a petri dish, if you will, for growing close ties—where their faith becomes an aspect of their social experience and daily lives.

Creating small groups, however, isn't enough. When Warren asked people what they discussed in one another's living rooms, he discovered they talked about the Bible and prayed together for ten minutes, and then spent the rest of the time discussing kids or gossiping. Warren's goal, however, wasn't just to help people make new friends. It was to build a community of the faithful, to encourage people to accept the lessons of Christ, and to make faith a focus of their lives. His small groups had created tight bonds, but without leadership, they weren't much more than a coffee circle. They weren't fulfilling his religious expectations.

Warren thought back to McGavran, the author. McGavran's philosophy said that if you teach people to live with Christian habits, they'll act as Christians without requiring constant guidance and monitoring. Warren couldn't lead every single small group in person; he couldn't be there to make sure every conversation focused on Christ instead of the latest TV shows. But if he gave people new habits, he figured, he wouldn't need to. When people gathered, their instincts would be to discuss the Bible, to pray together, to embody their faith.

So Warren created a series of curriculums, used in church classes and small group discussions, which were explicitly designed to teach parishioners new habits.

"If you want to have Christ-like character, then you just develop the habits that Christ had," one of Saddleback's course manuals reads. "All of us are simply a bundle of habits. . . . Our goal is to help you replace some bad habits with some good habits that will help you grow in Christ's likeness." Every Saddleback member is asked to sign a "maturity covenant card" promising to adhere to three habits: daily quiet time for reflection and prayer, tithing 10 percent of their income, and membership in a small group. Giving everyone new habits has become a focus of the church.

"Once we do that, the responsibility for spiritual growth is no longer with me, it's with you. We've given you a recipe," Warren told me. "We don't have to guide you, because you're guiding yourself. These habits become a new self-identity, and, at that point, we just need to support you and get out of your way."

Warren's insight was that he could expand his church the same way Martin Luther King grew the boycott: by relying on the combination of strong and weak ties. Transforming his church into a movement, however—scaling it across twenty thousand parishioners and thousands of other pastors—required something more, something that made it self-perpetuating. Warren needed to teach people habits that caused them to live faithfully not because of their ties, but because it's who they are.

This is the third aspect of how social habits drive movements: For an idea to grow beyond a community, it must become self-propelling. And the surest way to achieve that is to give people new habits that help them figure out where to go on their own.

● ● ●

As the bus boycott expanded from a few days into a week, and then a month, and then two months, the commitment of Montgomery's black community began to wane.

The police commissioner, citing an ordinance that required taxicabs to charge a minimum fare, threatened to arrest cabbies who drove blacks to work at a discount. The boycott's leaders responded by signing up two hundred volunteers to participate in a carpool. Police started issuing tickets and harassing people at carpool meeting spots. Drivers began dropping out. "It became more and more difficult to catch a ride," King later wrote. "Complaints began to rise. From early morning to late at night my telephone rang and my doorbell was seldom silent. I began to have doubts about the ability of the Negro community to continue the struggle."

One night, while King was preaching at his church, an usher ran up with an urgent message. A bomb had exploded at King's house while his wife and infant daughter were inside. King rushed home and was greeted by a crowd of several hundred blacks as well as the mayor and chief of police. His family had not been injured, but the front windows of his home were shattered and there was a crater in his porch. If anyone had been in the front rooms of the house when the bomb went off, they could have been killed.

As King surveyed the damage, more and more blacks arrived. Policemen started telling the crowds to disperse. Someone shoved a cop. A bottle flew through the air. One of the policemen swung a baton. The police chief, who months earlier had publicly declared his support for the racist White Citizens' Council, pulled King aside and asked him to do something—anything—to stop a riot from breaking out.

King walked to his porch.

"Don't do anything panicky," he shouted to the crowd. "Don't get your weapons. He who lives by the sword shall perish by the sword."

The crowd grew still.

"We must love our white brothers, no matter what they do to us," King said. "We must make them know that we love them. Jesus still cries out in words that echo across the centuries: 'Love your enemies; bless them that curse you; pray for them that despitefully use you.'"

It was the message of nonviolence that King had been increasingly preaching for weeks. Its theme, which drew on the writings of Gandhi and Jesus's sermons, was in many ways an argument listeners hadn't heard in this context before, a plea for nonviolent activism, overwhelming love and forgiveness of their attackers, and a promise that it would bring victory. For years, the civil rights movement had been kept alive by couching itself in the language of battles and struggles. There were contests and setbacks, triumphs and defeats that required everyone to recommit to the fight.

King gave people a new lens. This wasn't a war, he said. It was an embrace.

Equally important, King cast the boycott in a new and different light. This was not just about equality on buses, King said; it was part of God's plan, the same destiny that had ended British colonialism in India and slavery in the United States, and that had caused Christ to die on the cross so that he could take away our sins. It was the newest stage in a movement that had started centuries earlier. And as such, it required new responses, different strategies and behaviors. It needed participants to offer the other cheek. People could show their allegiance by adopting the new habits King was evangelizing about.

"We must meet hate with love," King told the crowd the night of the bombing. "If I am stopped, our work will not stop. For what we are doing is right. What we are doing is just. And God is with us."

When King was done speaking, the crowd quietly walked home.

"If it hadn't been for that nigger preacher," one white policeman later said, "we'd all be dead."

The next week, two dozen new drivers signed up for the carpool. The phone calls to King's home slowed. People began self-organizing, taking leadership of the boycott, propelling the movement. When more bombs exploded on the lawns of other boycott organizers, the same pattern played out. Montgomery's blacks showed up en masse, bore witness without violence or confrontation, and then went home.

It wasn't just in response to violence that this self-directed unity became visible. The churches started holding mass meetings every week—sometimes every night. "They were kind of like Dr. King's speech after the bombing—they took Christian teachings and made them political," Taylor Branch told me. "A movement is a saga. For it to work, everyone's identity has to change. People in Montgomery had to learn a new way to act."

Much like Alcoholics Anonymous—which draws power from group meetings where addicts learn new habits and start to believe by watching others demonstrate their faith—so Montgomery's citizens learned in mass meetings new behaviors that expanded the movement. "People went to see how other people were handling it," said Branch. "You start to see yourself as part of a vast social enterprise, and after a while, you really believe you are."

●●●

When the Montgomery police resorted to mass arrests to stop the boycott three months after it started, the community embraced the oppression. When ninety people were indicted by a grand jury, almost all of them rushed to the courthouse to present themselves for arrest. Some people went to the sheriff's office to see if their names were on the list and were "disappointed when they were not," King later wrote. "A once fear-ridden people had been transformed."

In future years, as the movement spread and there were waves of killings and attacks, arrests and beatings, the protesters—rather than fighting back, retreating, or using tactics that in the years before Montgomery had been activist mainstays—simply stood their ground and told white vigilantes that they were ready to forgive them when their hatred had ceased.

"Instead of stopping the movement, the opposition's tactics had only served to give it greater momentum, and to draw us closer together," King wrote. "They thought they were dealing with a group who could be cajoled or forced to do whatever the white man wanted them to do. They were not aware that they were dealing with Negroes who had been freed from fear."

There are, of course, numerous and complex reasons why the Montgomery bus boycott succeeded and why it became the spark for a movement that would spread across the South. But one critical factor is this third aspect of social habits. Embedded within King's philosophy was a set of new behaviors that converted participants from followers into self-directing leaders. These are not habits as we conventionally think about them. However, when King recast Montgomery's struggle by giving protesters a new sense of self-identity, the protest became a movement fueled by people who were acting because they had taken ownership of a historic event. And that social pattern, over time, became automatic and expanded to other places and groups of students and protesters whom King never met, but who could take on leadership of the movement simply by watching how its participants habitually behaved.

On June 5, 1956, a panel of federal judges ruled that Montgomery's bus segregation law violated the Constitution. The city appealed to the U.S. Supreme Court and on December 17, more than a year after Parks was arrested, the highest court rejected the final appeal. Three days later, city officials received the order: The buses had to be integrated.

The next morning, at 5:55 A.M., King, E. D. Nixon, Ralph Abernathy, and others climbed on board a city bus for the first time in more than twelve months, and sat in the front.

"I believe you are Reverend King, aren't you?" asked the white driver.

"Yes, I am."

"We are very glad to have you this morning," the driver said.

Later, NAACP attorney and future Supreme Court justice Thurgood Marshall would claim that the boycott had little to do with ending bus segregation in Montgomery. It was the Supreme Court, not capitulation by either side, that changed the law.

"All that walking for nothing," Marshall said. "They could just as well have waited while the bus case went up through the courts, without all the work and worry of the boycott."

Marshall, however, was wrong in one important respect. The Montgomery bus boycott helped birth a new set of social habits that quickly spread to Greensboro, North Carolina; Selma, Alabama; and Little Rock, Arkansas. The civil rights movement became a wave of sit-ins and peaceful demonstrations, even as participants were violently beaten. By the early 1960s, it had moved to Florida, California,

Washington, D.C., and the halls of Congress. When President Lyndon Johnson signed the Civil Rights Act of 1964—which outlawed all forms of segregation as well as discrimination against minorities and women—he equated the civil rights activists to the nation's founders, a comparison that, a decade earlier, would have been political suicide. "One hundred and eighty-eight years ago this week, a small band of valiant men began a long struggle for freedom," he told television cameras. "Now our generation of Americans has been called on to continue the unending search for justice within our own borders."

Movements don't emerge because everyone suddenly decides to face the same direction at once. They rely on social patterns that begin as the habits of friendship, grow through the habits of communities, and are sustained by new habits that change participants' sense of self.

King saw the power of these habits as early as Montgomery. "I cannot close without giving just a word of caution," he told a packed church on the night he called off the boycott. There was still almost a decade of protest ahead of him, but the end was in sight. "As we go back to the buses let us be loving enough to turn an enemy into a friend. We must now move from protest to reconciliation. . . . . With this dedication we will be able to emerge from the bleak and desolate midnight of man's inhumanity to man to the bright and glittering daybreak of freedom and justice."

9

THE NEUROLOGY OF FREE WILL
Are We Responsible for Our Habits?

I.

The morning the trouble began—years before she realized there was even trouble in the first place—Angie Bachmann was sitting at home, staring at the television, so bored that she was giving serious thought to reorganizing the silverware drawer.

Her youngest daughter had started kindergarten a few weeks earlier and her two older daughters were in middle school, their lives filled with friends and activities and gossip their mother couldn't possibly understand. Her husband, a land surveyor, often left for work at eight and didn't get home until six. The house was empty except for Bachmann. It was the first time in almost two decades—since she had gotten married at nineteen and pregnant by twenty, and her days had become crowded with packing school lunches, playing princess, and running a family shuttle service—that she felt genuinely alone. In high school, her friends told her she should become a model—she had been that pretty—but when she dropped out and then married a guitar player who eventually got a real job, she settled on being a mom instead. Now it was ten-thirty in the morning, her three daughters were gone, and Bachmann had resorted—again—to taping a piece of paper over the kitchen clock to stop herself from looking at it every three minutes.

She had no idea what to do next.

That day, she made a deal with herself: If she could make it until noon without going crazy or eating the cake in the fridge, she would leave the house and do something fun. She spent the next ninety minutes trying to figure out what exactly that would be. When the clock hit twelve o'clock, she put on some makeup and a nice dress and drove to a riverboat casino about twenty minutes away from her house. Even at noon on a Thursday, the casino was filled with people doing things besides watching soap operas and folding the laundry. There was a band playing near the entrance. A woman was handing out free cocktails. Bachmann ate shrimp from a buffet. The whole experience felt luxurious, like playing hooky. She made her way to a blackjack table where a dealer patiently explained the rules. When her forty dollars of chips were gone, she glanced at her watch and saw two hours had flown by and she needed to hurry home to pick up her youngest daughter. That night at dinner, for the first time in a month, she had something to talk about besides outguessing a contestant on *The Price Is Right*.

Angie Bachmann's father was a truck driver who had remade himself, midlife, into a semi-famous songwriter. Her brother had become a songwriter, too, and had won awards. Bachmann, on the other hand, was often introduced by her parents as "the one who became a mom."

"I always felt like the untalented one," she told me. "I think I'm smart, and I know I was a good mom. But there wasn't a lot I could point to and say, that's why I'm special."

After that first trip to the casino, Bachmann started going to the riverboat once a week, on Friday afternoons. It was a reward for making it through empty days, keeping the house clean, staying sane. She knew gambling could lead to trouble, so she set strict rules for herself. No more than one hour at the blackjack table per trip, and she only gambled what was in her wallet. "I considered it kind of like a job," she told me. "I never left the house before noon, and I was always home in time to pick up my daughter. I was very disciplined."

And she got good. At first, she could hardly make her money last an hour. Within six months, however, she had picked up enough tricks that she adjusted her rules to allow for two- or three-hour shifts, and she would still have cash in her pocket when she walked away. One afternoon, she sat down at the blackjack table with $80 in her purse and left with $530—enough to buy groceries, pay the phone bill, and put a bit in the rainy day fund. By then, the company that owned the casino—Harrah's Entertainment—was sending her coupons for free buffets. She would treat the family to dinner on Saturday nights.

The state where Bachmann was gambling, Iowa, had legalized gambling only a few years earlier. Prior to 1989, the state's lawmakers worried that the temptations of cards and dice might be difficult for some citizens to resist. It was a concern as old as the nation itself. Gambling "is the child of avarice, the brother of iniquity and the father of mischief," George Washington wrote in 1783. "This is a vice which is productive of every possible evil. . . . In a word, few gain by this abominable practice, while thousands are injured." Protecting people from their bad habits—in fact, defining which habits should be considered "bad" in the first place—is a prerogative lawmakers have eagerly seized. Prostitution, gambling, liquor sales on the Sabbath, pornography, usurious loans, sexual relations outside of marriage (or, if your tastes are unusual, within marriage), are all habits that various legislatures have regulated, outlawed, or tried to discourage with strict (and often ineffective) laws.

When Iowa legalized casinos, lawmakers were sufficiently concerned that they limited the activity to riverboats and mandated that no one could wager more than $5 per bet, with a maximum loss of $200 per person per cruise. Within a few years, however, after some of the state's casinos moved to Mississippi where no-limit gaming was allowed, the Iowa legislature lifted those restrictions. In 2010, the state's coffers swelled by more than $269 million from taxes on gambling.

● ● ●

In 2000, Angie Bachmann's parents, both longtime smokers, started showing signs of lung disease. She began flying to Tennessee to see them every other week, buying groceries and helping to cook dinner. When she came back home to her husband and daughters, the stretches seemed even lonelier now. Sometimes, the house was empty all day long; it was as if, in her absence, her friends had forgotten to invite her to things and her family had figured out how to get by on their own.

Bachmann was worried about her parents, upset that her husband seemed more interested in his work than her anxieties, and resentful of her kids who didn't realize she needed them now, after all the sacrifices she had made while they were growing up. But whenever she hit the casino, those tensions would float away. She started going a couple times a week when she wasn't visiting her parents, and then every Monday, Wednesday, and Friday. She still had rules—but she'd been gambling for years by now, and knew the axioms that serious players lived by. She never put down less than $25 a hand and always played two hands at once. "You have better odds at a higher limit table than at a lower limit table," she told me. "You have to be able to play through the rough patches until your luck turns. I've seen people walk in with $150 and win $10,000. I knew I could do this if I followed my rules. I was in control."* By then, she didn't have to think about whether to take another card or double her bet—she acted automatically, just as Eugene Pauly, the amnesiac, had eventually learned to always choose the right cardboard rectangle.

One day in 2000, Bachmann went home from the casino with $6,000—enough to pay rent for two months and wipe out the credit card bills that were piling up by the front door. Another time, she walked away with $2,000. Sometimes she lost, but that was part of the game. Smart gamblers knew you had to go down to go up. Eventually, Harrah's gave her a line of credit so she wouldn't have to carry so much cash. Other players sought her out and sat at her table because she knew what she was doing. At the buffet, the hosts would let her go to the front of the line. "I know how to play," she told me. "I know that sounds like somebody who's got a problem not recognizing their problem, but the only mistake I made was not quitting. There wasn't anything wrong with how I played."

Bachmann's rules gradually became more flexible as the size of her winnings and losses expanded. One day, she lost $800 in an hour, and then earned $1,200 in forty minutes. Then her luck turned again and she walked away down $4,000. Another time, she lost $3,500 in the morning, earned $5,000 by 1 P.M., and lost another $3,000 in the afternoon. The casino had records of how much she owed and what she'd earned; she'd stopped keeping track herself. Then, one month, she didn't have enough in her bank account for the electricity bill. She asked her parents for a small loan, and then another. She borrowed $2,000 one month, $2,500 the next. It wasn't a big deal; they had the money.

---

*It may seem irrational for anyone to believe they can beat the house in a casino. However, as regular gamblers know, it is possible to consistently win, particularly at games such as blackjack. Don Johnson of Bensalem, Pennsylvania, for instance, won a reported $15.1 million at blackjack over a six-month span starting in 2010. The house always wins in the aggregate because so many gamblers bet in a manner that doesn't maximize their odds, and most people do not have enough money to see themselves through losses. A gambler can consistently win over time, though, if he or she has memorized the complicated formulas and odds that guide how each hand should be played. Most players, however, don't have the discipline or mathematical skills to beat the house.

Bachmann never had problems with drinking or drugs or over-eating. She was a normal mom, with the same highs and lows as everyone else. So the compulsion she felt to gamble—the insistent pull that made her feel distracted or irritable on days when she didn't visit the casino, the way she found herself thinking about it all the time, the rush she felt on a good run—caught her completely off guard. It was a new sensation, so unexpected that she hardly knew it was a problem until it had taken hold of her life. In retrospect, it seemed like there had been no dividing line. One day it was fun, and the next it was uncontrollable.

By 2001, she was going to the casino every day. She went whenever she fought with her husband or felt unappreciated by her kids. At the tables she was numb and excited, all at once, and her anxieties grew so faint she couldn't hear them anymore. The high of winning was so immediate. The pain of losing passed so fast.

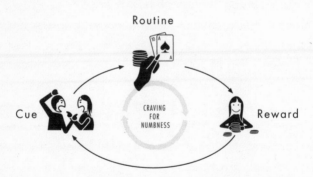

Routine

Cue

CRAVING FOR NUMBNESS

Reward

"You want to be a big shot," her mother told her when Bachmann called to borrow more money. "You keep gambling because you want the attention."

That wasn't it, though. "I just wanted to feel good at something," she said to me. "This was the only thing I'd ever done where it seemed like I had a skill."

By the summer of 2001, Bachmann's debts to Harrah's hit $20,000. She had been keeping the losses secret from her husband, but when her mother finally cut off the stipends, she broke down and confessed. They hired a bankruptcy attorney, cut up her credit cards, and sat at the kitchen table to write out a plan for a more austere, responsible life. She took her dresses to a used clothing store and withstood the humiliation of a nineteen-year-old turning down almost all of them because, she said, they were out of style.

Eventually, it started to feel like the worst was over. Finally, she thought, the compulsion was gone.

But, of course, it wasn't even close to the end. Years later, after she had lost everything and had ruined her life and her husband's, after she had thrown away hundreds of thousands of dollars and her lawyer had argued before the state's highest court that Angie Bachmann gambled not by choice, but out of habit, and thus shouldn't bear culpability for her losses, after she had become an object of scorn on the Internet, where people compared her to Jeffrey Dahmer and parents who abuse their kids, she would wonder: How much responsibility do I actually bear?

"I honestly believe anyone in my shoes would have done the same things," Bachmann told me.

II.

On a July morning in 2008, a desperate man vacationing along the west coast of Wales picked up the phone and called an emergency operator.

"I think I've killed my wife," he said. "Oh my God. I thought someone had broken in. I was fighting with those boys but it was Christine. I must have been dreaming or something. What have I done? What have I done?"

Ten minutes later, police officers arrived to find Brian Thomas crying next to his camper van. The previous night, he explained, he and his wife had been sleeping in the van when young men racing around the parking lot had awoken them. They moved their camper to the edge of the lot and went back to sleep. Then, a few hours later,

Thomas woke to find a man in jeans and a black fleece—one of the racers, he thought—lying on top of his wife. He screamed at the man, grabbed him by the throat, and tried to pull him off. It was as if he was reacting automatically, he told the police. The more the man struggled, the harder Thomas squeezed. The man scratched at Thomas's arm and tried to fight back, but Thomas choked, tighter and tighter, and eventually the man stopped moving. Then, Thomas realized it wasn't a man in his hands, but his wife. He dropped her body and began gently nudging her shoulder, trying to wake her, asking if she was all right. It was too late.

"I thought somebody had broken in and I strangled her," Thomas told the police, sobbing. "She's my world."

For the next ten months, as Thomas sat in prison awaiting trial, a portrait of the murderer emerged. As a child, Thomas had started sleepwalking, sometimes multiple times each night. He would get out of bed, walk around the house and play with toys or fix himself something to eat and, the next morning, remember nothing about what he had done. It became a family joke. Once a week, it seemed, he would wander into the yard or someone else's room, all while asleep. It was a habit, his mother would explain when neighbors asked why her son was walking across their lawns, barefoot and in his pajamas. As he grew older, he would wake up with cuts on his feet and no memories of where they had come from. He once swam in a canal without waking. After he married, his wife grew so concerned about the possibility that he might stumble out of the house and into traffic that she locked the door and slept with the keys under her pillow. Every night, the couple would crawl into bed and "have a kiss and a cuddle," Thomas later said, and then he would go to his own room and sleep in his own bed. Otherwise his restless tossing and turning, the shouting and grunting and occasional wanderings, would keep Christine up all night.

"Sleepwalking is a reminder that wake and sleep are not mutually exclusive," Mark Mahowald, a professor of neurology at the University of Minnesota and a pioneer in understanding sleep behaviors, told me. "The part of your brain that monitors your behavior is asleep, but the parts capable of very complex activities are awake. The problem is that there's nothing guiding the brain except for basic patterns, your most basic habits. You follow what exists in your head, because you're not capable of making a choice."

By law, the police had to prosecute Thomas for the murder. But all evidence seemed to indicate that he and his wife had a happy marriage prior to that awful night. There wasn't any history of abuse. They had two grown daughters and had recently booked a Mediterranean cruise to celebrate their fortieth wedding anniversary. Prosecutors asked a sleep specialist—Dr. Chris Idzikowski of the Edinburgh Sleep Centre—to examine Thomas and evaluate a theory: that he had been unconscious when he killed his wife. In two separate sessions, one in Idzikowski's laboratory and the other inside the prison, the researcher applied sensors all over Thomas's body and measured his brain waves, eye movement, chin and leg muscles, nasal airflow, respiratory effort, and oxygen levels while he slept.

Thomas wasn't the first person to argue that he had committed a crime while sleeping and thus, by extension, should not be held responsible for his deed. There's a long history of wrongdoers contending they aren't culpable due to "automatism," as sleepwalking and other unconscious behaviors are known. And in the past decade, as our understanding of the neurology of habits and free will has become more sophisticated, those defenses have become more compelling. Society, as embodied by our courts and juries, has agreed that some habits are so powerful that they overwhelm our capacity to make choices, and thus we're not responsible for what we do.

● ● ●

Sleepwalking is an odd outgrowth of a normal aspect of how our brains work while we slumber. Most of the time, as our bodies move in and out of different phases of rest, our most primitive neurological structure—the brain stem—paralyzes our limbs and nervous system, allowing our brains to experience dreams without our bodies moving. Usually, people can make the transition in and out of paralysis multiple times each night without any problems. Within neurology, it's known as the "switch."

Some people's brains, though, experience switching errors. They go into incomplete paralysis as they sleep, and their bodies are active while they dream or pass between sleep phases. This is the root cause of sleepwalking and for the majority of sufferers, it is an annoying but benign problem. Someone might dream about eating a cake, for instance, and the next morning find a ravaged box of doughnuts in the kitchen. Someone will dream about going to the bathroom, and later discover a wet spot in the hall. Sleepwalkers can behave in complex ways—for instance, they can open their eyes, see, move around, and drive a car or cook a meal—all while essentially unconscious, because the parts of their brain associated with seeing, walking, driving, and cooking can function while they are asleep without input from the brain's more advanced regions, such as the prefrontal cortex. Sleepwalkers have been known to boil water and make tea. One operated a motorboat. Another turned on an electric saw and started feeding in pieces of wood before going back to bed. But in general, sleepwalkers will not do things that are dangerous to themselves or others. Even asleep, there's an instinct to avoid peril.

However, as scientists have examined the brains of sleepwalkers, they've found a distinction between *sleepwalking*—in which people might leave their beds and start acting out their dreams or other mild impulses—and something called *sleep terrors*. When a sleep terror occurs, the activity inside people's brains is markedly different from when they are awake, semi-conscious, or even sleepwalking. People in the midst of sleep terrors seem to be in the grip of terrible anxieties, but are not dreaming in the normal sense of the word. Their brains shut down except for the most primitive neurological regions, which include what are known as "central pattern generators." These areas of the brain are the same ones studied by Dr. Larry Squire and the scientists at MIT, who found the neurological machinery of the habit loop. To a neurologist, in fact, a brain experiencing a sleep terror looks very similar to a brain following a habit.

The behaviors of people in the grip of sleep terrors *are* habits, though of the most primal kind. The "central pattern generators" at work during a sleep terror are where such behavioral patterns as walking, breathing, flinching from a loud noise, or fighting an attacker come from. We don't usually think about these behaviors as habits, but that's what they are: automatic behaviors so ingrained in our neurology that, studies show, they can occur with almost no input from the higher regions of the brain.

However, these habits, when they occur during sleep terrors, are different in one critical respect: Because sleep deactivates the prefrontal cortex and other high cognition areas, when a sleep terror habit is triggered, there is no possibility of conscious intervention. If the fight-or-flight habit is cued by a sleep terror, there is no chance that someone can override it through logic or reason.

"People with sleep terrors aren't dreaming in the normal sense," said Mahowald, the neurologist. "There's no complex plots like you and I remember from a nightmare. If they remember anything afterward, it's just an image or emotions—impending doom, horrible fear, the need to defend themselves or someone else.

"Those emotions are really powerful, though. They are some of the most basic cues for all kinds of behaviors we've learned throughout our lives. Responding to a threat by running away or defending ourselves is something everyone has practiced since they were babies. And when those emotions occur, and there's no chance for the higher brain to put things in context, we react the way our deepest habits tell us to. We run or fight or follow whatever behavioral pattern is easiest for our brains to latch on to."

When someone in the midst of a sleep terror starts feeling threatened or sexually aroused—two of the most common sleep terror experiences—they react by following the habits associated with those stimuli. People experiencing sleep terrors have jumped off of tall roofs because they believed they were fleeing from attackers. They have killed their own babies because, they believed, they were fighting wild animals. They have raped their spouses, even as their victims begged them to stop, because once the sleepers' arousal began, they followed the ingrained habit to satisfy the urge. Sleepwalking seems to allow some choice, some participation by our higher brains that tell us to stay away from the edge of the roof. Someone in the grip of a sleep terror, however, simply follows the habit loop no matter where it leads.

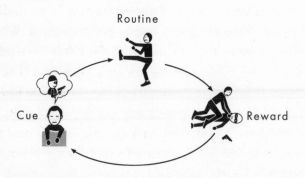

Routine

Cue

Reward

Some scientists suspect sleep terrors might be genetic; others say diseases such as Parkinson's make them more likely. Their causes aren't well understood, but for a number of people, sleep terrors involve violent impulses. "Violence related to sleep terrors appears to be a reaction to a concrete, frightening image that the individual can subsequently describe," a group of Swiss researchers wrote in 2009. Among people suffering one type of sleep dysfunction, "attempted assault of sleep partners has been reported to occur in 64% of cases, with injuries in 3%."

In both the United States and the United Kingdom, there is a history of murderers arguing that sleep terrors caused them to commit crimes they would have never consciously carried out. Four years before Thomas was arrested, for instance, a man named Jules Lowe was found not guilty of murdering his eighty-three-year-old father after claiming that the attack occurred during a sleep terror. Prosecutors argued it was "far-fetched in the extreme" to believe that Lowe was asleep while he punched, kicked, and stamped his father for more than twenty minutes, leaving him with over ninety injuries. The jury disagreed and set him free. In September 2008, thirty-three-year-old Donna Sheppard-Saunders nearly suffocated her mother by holding a pillow over her face for thirty seconds. She was later acquitted of attempted murder by arguing that she had acted while asleep. In 2009, a British soldier admitted to raping a teenage girl, but said he was asleep and unconscious while he undressed himself, pulled down her pants, and began having sex. When he woke, mid-rape, he apologized and called the police. "I've just sort of committed a crime," he told the emergency operator. "I honestly don't know what happened. I woke up on top of her." He had a history of suffering from sleep terrors and was found not guilty. More than 150 murderers and rapists have escaped punishment in the past century using the automatism defense. Judges and juries, acting on behalf of society, have said that since the criminals didn't *choose* to commit their crimes—since they didn't consciously participate in the violence—they shouldn't bear the blame.

For Brian Thomas, it also looked like a situation where a sleep disorder, rather than a murderous impulse, was at fault. "I'll never forgive myself, ever," he told one of the prosecutors. "Why did I do it?"

● ● ●

After Dr. Idzikowski, the sleep specialist, observed Thomas in his laboratory, he submitted his findings: Thomas was asleep when he killed his wife. He hadn't consciously committed a crime.

As the trial started, prosecutors presented their evidence to the jury. Thomas had admitted to murdering his wife, they told jurors. He knew he had a history of sleepwalking. His failure to take precautions while on vacation, they said, made him responsible for his crime.

But as arguments proceeded, it became clear prosecutors were fighting an uphill battle. Thomas's lawyer argued that his client hadn't meant to kill his wife—in fact, he wasn't even in control of his own actions that night. Instead, he was reacting automatically to a perceived threat. He was following a habit almost as old as our species: the instinct to fight an attacker and protect a loved one. Once the most primitive parts of his brain were exposed to a cue—someone strangling his wife—his habit took over and he fought back, with no chance of his higher cognition interceding. Thomas was guilty of nothing more than being human, the lawyer argued, and reacting in the way his neurology—and most primitive habits—forced him to behave.

Even the prosecution's own witnesses seemed to bolster the defense. Though Thomas had known he was capable of sleepwalking, the prosecution's own psychiatrists said, there was nothing to suggest to him that it was therefore foreseeable he might kill. He had never attacked anyone in his sleep before. He had never previously harmed his wife.

When the prosecution's chief psychiatrist took the stand, Thomas's lawyer began his cross-examination.

Did it seem fair that Thomas should be found guilty for an act he could not know was going to occur?

In her opinion, said Dr. Caroline Jacob, Thomas could not have reasonably anticipated his crime. And if he was convicted and sentenced to Broadmoor Hospital, where some of Britain's most dangerous and mentally ill criminals were housed, well, "he does not belong there."

The next morning, the head prosecutor addressed the jury.

"At the time of the killing the defendant was asleep and his mind had no control over what his body was doing," he said. "We have reached the conclusion that the public interest would no longer be served by continuing to seek a special verdict from you. We therefore offer no further evidence and invite you to return a straight not guilty verdict." The jury did so.

Before Thomas was set free, the judge told him, "You are a decent man and a devoted husband. I strongly suspect you may well be feeling a sense of guilt. In the eyes of the law you bear no responsibility. You are discharged."

It seems like a fair outcome. After all, Thomas was obviously devastated by his crime. He had no idea what he was doing when he acted—he was simply following a habit, and his capacity for decision making was, in effect, incapacitated. Thomas is the most sympathetic murderer conceivable, someone so close to being a victim himself that when the trial ended, the judge tried to console him.

Yet many of those same excuses can be made for Angie Bachmann, the gambler. She was also devastated by her actions. She would later say she carries a deep sense of guilt. And as it turns out, she was also following deeply ingrained habits that made it increasingly difficult for decision making to intervene.

But in the eyes of the law Bachmann is responsible for her habits, and Thomas isn't. Is it right that Bachmann, a gambler, is guiltier than Thomas, a murderer? What does that tell us about the ethics of habit and choice?

III.

Three years after Angie Bachmann declared bankruptcy, her father passed away. She'd spent the previous half decade flying between her home and her parents' house, tending to them as they became increasingly ill. His death was a blow. Then, two months later, her mother died.

"My entire world disintegrated," she said. "I would wake up every morning, and for a second forget they had passed, and then it would rush in that they were gone and I'd feel like someone was standing on my chest. I couldn't think about anything else. I didn't know what to do when I got out of bed."

When their wills were read, Bachmann learned she had inherited almost $1 million.

She used $275,000 to buy her family a new home in Tennessee, near where her mother and father had lived, and spent a bit more to move her grown daughters nearby so everyone was close. Casino gambling wasn't legal in Tennessee, and "I didn't want to fall back into bad patterns," she told me. "I wanted to live away from anything that reminded me of feeling out of control." She changed her phone numbers and didn't tell the casinos her new address. It felt safer that way.

Then one night, driving through her old hometown with her husband, picking up the last of their furniture from her previous home, she started thinking about her parents. How would she manage without them? Why hadn't she been a better daughter? She began hyperventilating. It felt like the beginning of a panic attack. It had been years since she had gambled, but in that moment she felt like she needed to find something to take her mind off the pain. She looked at her husband. She was desperate. This was a one-time thing.

"Let's go to the casino," she said.

When they walked in, one of the managers recognized her from when she was a regular and invited them into the players' lounge. He asked how she had been, and it all came tumbling out: her parents' passing and how hard it had hit her, how exhausted she was all the time, how she felt like she was on the verge of a breakdown. The manager was a good listener. It felt so good to finally say everything she had been thinking and be told that it was normal to feel this way.

Then she sat down at a blackjack table and played for three hours.

For the first time in months, the anxiety faded into background noise. She knew how to do this. She went blank. She lost a few thousand dollars.

Harrah's Entertainment—the company that owned the casino—was known within the gaming industry for the sophistication of its customer-tracking systems. At the core of that system were computer programs much like those Andrew Pole created at Target, predictive algorithms that studied gamblers' habits and tried to figure out how to persuade them to spend more. The company assigned players a "predicted lifetime value," and software built calendars that anticipated how often they would visit and how much they would spend. The company tracked customers through loyalty cards and mailed out coupons for free meals and cash vouchers; telemarketers called people at home to ask where they had been. Casino employees were trained to encourage visitors to discuss their lives, in the hopes they might reveal information that could be used to predict how much they had to gamble with. One Harrah's executive called this approach "Pavlovian marketing." The company ran thousands of tests each year to perfect their methods. Customer tracking had increased the company's profits by billions of dollars, and was so precise they could track a gambler's spending to the cent and minute.*

Harrah's, of course, was well aware that Bachmann had declared bankruptcy a few years earlier and had walked away from $20,000 in gambling debts. But soon after her conversation with the casino manager, she began receiving phone calls with offers of free limos that would take her to casinos in Mississippi. They offered to fly her and her husband to Lake Tahoe, put them in a suite, and give them tickets to an Eagles concert. "I said my daughter has to come, and she wants to bring a friend," Bachmann said. No problem, the company replied. Everyone's airfare and rooms were free. At the concert,

---

*Harrah's—now known as Caesars Entertainment—disputes some of Bachmann's allegations. Their comments can be found in the notes.

she sat in the front row. Harrah's gave her $10,000 to play with, compliments of the house.

The offers kept coming. Every week another casino called, asking if she wanted a limo, entry to shows, plane tickets. Bachmann resisted at first, but eventually she started saying yes each time an invitation arrived. When a family friend mentioned that she wanted to get married in Las Vegas, Bachmann made a phone call and the next weekend they were in the Palazzo. "Not that many people even know it exists," she told me. "I've called and asked about it, and the operator said it's too exclusive to give out information over the phone. The room was like something out of a movie. It had six bedrooms and a deck and private hot tub for each room. I had a butler."

When she got to the casinos, her gambling habits took over almost as soon as she walked in. She would often play for hours at a stretch. She started small at first, using only the casino's money. Then the numbers got larger, and she would replenish her chips with withdrawals from the ATM. It didn't seem to her like there was a problem. Eventually she was playing $200 to $300 per hand, two hands at a time, sometimes for a dozen hours at a time. One night, she won $60,000. Twice she walked away up $40,000. One time she went to Vegas with $100,000 in her bag and came home with nothing. It didn't really change her lifestyle. Her bank account was still so large that she never had to think about money. That's why her parents had left her the inheritance in the first place: so she could enjoy herself.

She would try to slow down, but the casino's appeals became more insistent. "One host told me that he would get fired if I didn't come in that weekend," she said. "They would say, 'We sent you to this concert and we gave you this nice room, and you haven't been gambling that much lately.' Well, they *did* do those nice things for me."

In 2005, her husband's grandmother died and the family went back to her old hometown for the funeral. She went to the casino the night before the service to clear her head and get mentally prepared for all the activity the next day. Over a span of twelve hours, she lost $250,000. At the time, it was almost as if the scale of the loss didn't register. When she thought about it afterward—*a quarter of a million dollars gone*—it didn't seem real. She had lied to herself about so much already: that her marriage was happy when she and her husband sometimes went days without really speaking; that her friends were close when she knew they appeared for Vegas trips and were gone when it was over; that she was a good mom when she saw her daughters making the same mistakes she had made, getting pregnant too early; that her parents would have been pleased to see their money thrown away this way. It felt like there were only two choices: continue lying to herself or admit that she had dishonored everything her mother and father had worked so hard to earn.

A quarter of a million dollars. She didn't tell her husband. "I concentrated on something new whenever that night popped into my mind," she said.

Soon, though, the losses were too big to ignore. Some nights, after her husband was asleep, Bachmann would crawl out of bed, sit at the kitchen table, and scribble out figures, trying to make sense of how much was gone. The depression that had started after her parents' death seemed to be getting deeper. She felt so tired all the time.

And Harrah's kept calling.

"This desperation starts once you realize how much you've lost, and then you feel like you can't stop because you've got to win it back," she said. "Sometimes I'd start feeling jumpy, like I couldn't think straight, and I'd know that if I pretended I might take another trip soon, it would calm me down. Then they would call and I'd say yes because it was so easy to give in. I really believed I might win it back. I'd won before. If you couldn't win, then gambling wouldn't be legal, right?"

    ●●●

In 2010, a cognitive neuroscientist named Reza Habib asked twenty-two people to lie inside an MRI and watch a slot machine spin around and around. Half of the participants were "pathological gamblers"—people who had lied to their families about their gambling, missed work to gamble, or had bounced checks at a casino—while the other half were people who gambled socially but didn't exhibit any problematic behaviors. Everyone was placed on their backs inside a narrow tube and told to watch wheels of lucky 7s, apples, and gold bars spin across a video screen. The slot machine was programmed to deliver three outcomes: a win, a loss, and a "near miss," in which the slots almost matched up but, at the last moment, failed to align. None of the participants won or lost any money. All they had to do was watch the screen as the MRI recorded their neurological activity.

"We were particularly interested in looking at the brain systems involved in habits and addictions," Habib told me. "What we found was that, neurologically speaking, pathological gamblers got more excited about winning. When the symbols lined up, even though they didn't actually win any money, the areas in their brains related to emotion and reward were much more active than in non-pathological gamblers.

"But what was really interesting were the *near misses*. To pathological gamblers, near misses looked like wins. Their brains reacted almost the same way. But to a nonpathological gambler, a near miss was like a loss. People without a gambling problem were better at recognizing that a near miss means you still lose."

Two groups saw the exact same event, but from a neurological perspective, they viewed it differently. People with gambling problems got a mental high from the near misses—which, Habib hypothesizes, is probably why they gamble for so much longer than everyone else: because the near miss triggers those habits that prompt them to put down another bet. The nonproblem gamblers, when they saw a near miss, got a dose of apprehension that triggered a different habit, the one that says *I should quit before it gets worse.*

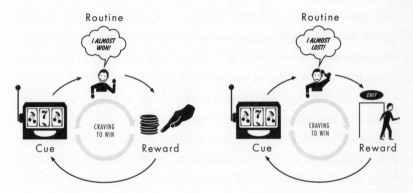

It's unclear if problem gamblers' brains are different because they are born that way or if sustained exposure to slot machines, online poker, and casinos can change how the brain functions. What is clear is that real neurological differences impact how pathological gamblers process information—which helps explain why Angie Bachmann lost control every time she walked into a casino. Gaming companies are well aware of this tendency, of course, which is why in the past decades, slot machines have been reprogrammed to deliver a more constant supply of near wins.* Gamblers who keep betting after near wins are what make casinos, racetracks, and state lotteries so profitable. "Adding a near miss to a lottery is like pouring jet fuel on a fire," said a state lottery consultant who spoke to me on the condition of anonymity. "You want to know why sales have exploded? Every other scratch-off ticket is designed to make you feel like you almost won."

*In the late 1990s, one of the largest slot machine manufacturers hired a former video game executive to help them design new slots. That executive's insight was to program machines to deliver more near wins. Now, almost every slot contains numerous twists—such as free spins and sounds that erupt when icons almost align—as well as small payouts that make players feel like they are winning when, in truth, they are putting in more money than they are getting back. "No other form of gambling manipulates the human mind as beautifully as these machines," an addictive-disorder researcher at the University of Connecticut School of Medicine told a *New York Times* reporter in 2004.

The areas of the brain that Habib scrutinized in his experiment—the basal ganglia and the brain stem—are the same regions where habits reside (as well as where behaviors related to sleep terrors start). In the past decade, as new classes of pharmaceuticals have emerged that target that region—such as medications for Parkinson's disease—we've learned a great deal about how sensitive some habits can be to outside stimulation. Class action lawsuits in the United States, Australia, and Canada have been filed against drug manufacturers, alleging that pharmaceuticals caused patients to compulsively bet, eat, shop, and masturbate by targeting the circuitry involved in the habit loop. In 2008, a federal jury in Minnesota awarded a patient $8.2 million in a lawsuit against a drug company after the man claimed that his medication had caused him to gamble away more than $250,000. Hundreds of similar cases are pending.

"In those cases, we can definitively say that patients have no control over their obsessions, because we can point to a drug that impacts their neurochemistry," said Habib. "But when we look at the brains of people who are obsessive gamblers, they look very similar—except they can't blame it on a medication. They tell researchers they don't want to gamble, but they can't resist the cravings. So why do we say that those gamblers are in control of their actions and the Parkinson's patients aren't?"

● ● ●

On March 18, 2006, Angie Bachmann flew to a casino at Harrah's invitation. By then, her bank account was almost empty. When she tried to calculate how much she had lost over her lifetime, she put the figure at about $900,000. She had told Harrah's that she was almost broke, but the man on the phone said to come anyway. They would give her a line of credit, he said.

"It felt like I couldn't say no, like whenever they dangled the smallest temptation in front of me, my brain would shut off. I know that sounds like an excuse, but they always promised it would be different this time, and I knew no matter how much I fought against it, I was eventually going to give in."

She brought the last of her money with her. She started playing $400 a hand, two hands at a time. If she could get up a little bit, she told herself, just $100,000, she could quit and have something to give her kids. Her husband joined her for a while, but at midnight he went to bed. Around 2 A.M., the money she had come with was gone. A Harrah's employee gave her a promissory note to sign. Six times she signed for more cash, for a total of $125,000.

At about six in the morning, she hit a hot streak and her piles of chips began to grow. A crowd gathered. She did a quick tally: not quite enough to pay off the notes she had signed, but if she kept playing smart, she would come out on top, and then quit for good. She won five times in a row. She only needed to win $20,000 more to pull ahead. Then the dealer hit 21. Then he hit it again. A few hands later, he hit it a third time. By ten in the morning, all her chips were gone. She asked for more credit, but the casino said no.

Bachmann left the table dazed and walked to her suite. It felt like the floor was shaking. She trailed a hand along the wall so that if she fell, she'd know which way to lean. When she got to the room, her husband was waiting for her.

"It's all gone," she told him.

"Why don't you take a shower and go to bed?" he said. "It's okay. You've lost before."

"It's all gone," she said.

"What do you mean?"

"The money is gone," she said. "All of it."

"At least we still have the house," he said.

She didn't tell him that she'd taken out a line of credit on their home months earlier and had gambled it away.

IV.

Brian Thomas murdered his wife. Angie Bachmann squandered her inheritance. Is there a difference in how society should assign responsibility?

Thomas's lawyer argued that his client wasn't culpable for his wife's death because he acted unconsciously, automatically, his reaction cued by the belief that an intruder was attacking. He never *chose* to kill, his lawyer said, and so he shouldn't be held responsible for her death. By the same logic, Bachmann—as we know from Reza Habib's research on the brains of problem gamblers—was also driven by powerful cravings. She may have made a choice that first day when she got dressed up and decided to spend the afternoon in a casino, and perhaps in the weeks or months that followed. But years later, by the time she was losing $250,000 in a single night, after she was so desperate to fight the urges that she moved to a state where gambling wasn't legal, she was no longer making conscious decisions. "Historically, in neuroscience, we've said that people with brain damage lose some of their free will," said Habib. "But when a pathological gambler sees a casino, it seems very similar. It seems like they're acting without choice."

Thomas's lawyer argued, in a manner that everyone believed, that his client had made a terrible mistake and would carry the guilt of it for life. However, isn't it clear that Bachmann feels much the same way? "I feel so guilty, so ashamed of what I've done," she told me. "I feel like I've let everyone down. I know that I'll never be able to make up for this, no matter what I do."

That said, there is one critical distinction between the cases of Thomas and Bachmann: Thomas murdered an innocent person. He committed what has always been the gravest of crimes. Angie Bachmann lost money. The only victims were herself, her family, and a $27 billion company that loaned her $125,000.

Thomas was set free by society. Bachmann was held accountable for her deeds.

Ten months after Bachmann lost everything, Harrah's tried to collect from her bank. The promissory notes she signed bounced, and so Harrah's sued her, demanding Bachmann pay her debts and an additional $375,000 in penalties—a civil punishment, in effect, for committing a crime. She countersued, claiming that by extending her credit, free suites, and booze, Harrah's had preyed on someone they knew had no control over her habits. Her case went all the way to the state Supreme Court. Bachmann's lawyer—echoing the arguments that Thomas's attorney had made on the murderer's behalf—said that she shouldn't be held culpable because she had been reacting automatically to temptations that Harrah's put in front of her. Once the offers started rolling in, he argued, once she walked into the casino, her habits took over and it was impossible for her to control her behavior.

The justices, acting on behalf of society, said Bachmann was wrong. "There is no common law duty obligating a casino operator to refrain from attempting to entice or contact gamblers that it knows or should know are compulsive gamblers," the court wrote. The state had a "voluntary exclusion program" in which any person could ask for their name to be placed upon a list that required casinos to bar them from playing, and "the existence of the voluntary exclusion program suggests the legislature intended pathological gamblers to take personal responsibility to prevent and protect themselves against compulsive gambling," wrote Justice Robert Rucker.

Perhaps the difference in outcomes for Thomas and Bachmann is fair. After all, it's easier to sympathize with a devastated widower than a housewife who threw everything away.

Why is it easier, though? Why does it seem the bereaved husband is a victim, while the bankrupt gambler got her just deserts? Why do some habits seem like they should be so easy to control, while others seem out of reach?

More important, is it right to make a distinction in the first place?

"Some thinkers," Aristotle wrote in *Nicomachean Ethics,* "hold that it is by nature that people become good, others that it is by habit, and others that it is by instruction." For Aristotle, habits reigned supreme. The behaviors that occur unthinkingly are the evidence of our truest selves, he said. So "just as a piece of land has to be prepared beforehand if it is to nourish the seed, so the mind of the pupil has to be prepared in its habits if it is to enjoy and dislike the right things."

Habits are not as simple as they appear. As I've tried to demonstrate throughout this book, habits—even once they are rooted in our minds—aren't destiny. We can choose our habits, once we know how. Everything we know about habits, from neurologists studying amnesiacs and organizational experts remaking companies, is that any of them can be changed, if you understand how they function.

Hundreds of habits influence our days—they guide how we get dressed in the morning, talk to our kids, and fall asleep at night; they impact what we eat for lunch, how we do business, and whether we exercise or have a beer after work. Each of them has a different cue and offers a unique reward. Some are simple and others are complex, drawing upon emotional triggers and offering subtle neurochemical prizes. But every habit, no matter its complexity, is malleable. The most addicted alcoholics can become sober. The most dysfunctional companies can transform themselves. A high school dropout can become a successful manager.

However, to modify a habit, you must *decide* to change it. You must consciously accept the hard work of identifying the cues and rewards that drive the habits' routines, and find alternatives. You must know you have control and be self-conscious enough to use it—and every chapter in this book is devoted to illustrating a different aspect of why that control is real.

So though both Angie Bachmann and Brian Thomas made variations on the same claim—that they acted out of habit, that they had no control over their actions because those behaviors unfolded automatically—it seems fair that they should be treated differently. It is just that Angie Bachmann should be held accountable and that Brian Thomas should go free because Thomas never knew the patterns that drove him to kill existed in the first place—much less that he could master them. Bachmann, on the other hand, was aware of her habits. And once you know a habit exists, you have the responsibility to change it. If she had tried a bit harder, perhaps she could have reined them in. Others have done so, even in the face of greater temptations.

That, in some ways, is the point of this book. Perhaps a sleep-walking murderer can plausibly argue he wasn't aware of his habit, and so he doesn't bear responsibility for his crime. But almost all the other patterns that exist in most people's lives—how we eat and sleep and talk to our kids, how we unthinkingly spend our time, attention, and money—those *are* habits that we know exist. And once you understand that habits can change, you have the freedom—and the responsibility—to remake them. Once you understand that habits can be rebuilt, the power of habit becomes easier to grasp, and the only option left is to get to work.

●●●

"All our life," William James told us in the prologue, "so far as it has definite form, is but a mass of habits—practical, emotional, and intellectual—systematically organized for our weal or woe, and bearing us irresistibly toward our destiny, whatever the latter may be."

James, who died in 1910, hailed from an accomplished family. His father was a wealthy and prominent theologian. His brother, Henry, was a brilliant, successful writer whose novels are still studied today. William, into his thirties, was the unaccomplished one in the family. He was sick as a child. He wanted to become a painter, and then enrolled in medical school, then left to join an expedition up the Amazon River. Then he quit that, as well. He chastised himself in his diary for not being good at anything. What's more, he wasn't certain if he could get better. In medical school, he had visited a hospital for the insane and had seen a man hurling himself against a wall. The patient, a doctor explained, suffered from hallucinations. James didn't say that he often felt like he shared more in common with the patients than his fellow physicians.

"Today I about touched bottom, and perceive plainly that I must face the choice with open eyes," James wrote in his diary in 1870, when he was twenty-eight years old. "Shall I frankly throw the moral business overboard, as one unsuited to my innate aptitudes?"

Is suicide, in other words, a better choice?

Two months later, James made a decision. Before doing anything rash, he would conduct a yearlong experiment. He would spend twelve months believing that he had control over himself and his destiny, that he could become better, that he had the free will to change. There was no proof that it was true. But he would free himself to *believe*, all evidence to the contrary, that change was possible. "I think that yesterday was a crisis in my life," he wrote in his diary. Regarding his ability to change, "I will assume for the present—until next year—that it is no illusion. My first act of free will shall be to believe in free will."

Over the next year, he practiced every day. In his diary, he wrote as if his control over himself and his choices was never in question. He got married. He started teaching at Harvard. He began spending time with Oliver Wendell Holmes, Jr., who would go on to become a Supreme Court justice, and Charles Sanders Peirce, a pioneer in the study of semiotics, in a discussion group they called the Metaphysical Club. Two years after writing his diary entry, James sent a letter to the philosopher Charles Renouvier, who had expounded at length on free will. "I must not lose this opportunity of telling you of the admiration and gratitude which have been excited in me by the reading of your *Essais*," James wrote. "Thanks to you I possess for the first time an intelligible and reasonable conception of freedom. . . . I can say that through that philosophy I am beginning to experience a rebirth of the moral life; and I can assure you, sir, that this is no small thing."

Later, he would famously write that the will to believe is the most important ingredient in creating belief in change. And that one of the most important methods for creating that belief was habits. Habits, he noted, are what allow us to "do a thing with difficulty the first time, but soon do it more and more easily, and finally, with sufficient practice, do it semi-mechanically, or with hardly any consciousness at all." Once we choose who we want to be, people grow "to the way in which they have been exercised, just as a sheet of paper or a coat, once creased or folded, tends to fall forever afterward into the same identical folds."

If you believe you can change—if you make it a habit—the change becomes real. This is the real power of habit: the insight that your habits are what you choose them to be. Once that choice occurs—and becomes automatic—it's not only real, it starts to seem inevitable, the thing, as James wrote, that bears "us irresistibly toward our destiny, whatever the latter may be."

The way we habitually think of our surroundings and ourselves create the worlds that each of us inhabit. "There are these two young fish swimming along and they happen to meet an older fish swimming the other way, who nods at them and says 'Morning, boys. How's the water?'" the writer David Foster Wallace told a class of graduating college students in 2005. "And the two young fish swim on for a bit, and then eventually one of them looks over at the other and goes 'What the hell is water?'"

The water is habits, the unthinking choices and invisible decisions that surround us every day—and which, just by looking at them, become visible again.

Throughout his life, William James wrote about habits and their central role in creating happiness and success. He eventually devoted an entire chapter in his masterpiece *The Principles of Psychology* to the topic. Water, he said, is the most apt analogy for how a habit works. Water "hollows out for itself a channel, which grows broader and deeper; and, after having ceased to flow, it resumes, when it flows again, the path traced by itself before."

You now know how to redirect that path. You now have the power to swim.

APPENDIX

A Reader's Guide to Using These Ideas

The difficult thing about studying the science of habits is that most people, when they hear about this field of research, want to know the secret formula for quickly changing any habit. If scientists have discovered how these patterns work, then it stands to reason that they must have also found a recipe for rapid change, right?

If only it were that easy.

It's not that formulas don't exist. The problem is that there isn't one formula for changing habits. There are thousands.

Individuals and habits are all different, and so the specifics of diagnosing and changing the patterns in our lives differ from person to person and behavior to behavior. Giving up cigarettes is different from curbing overeating, which is different from changing how you communicate with your spouse, which is different from how you prioritize tasks at work. What's more, each person's habits are driven by different cravings.

As a result, this book doesn't contain one prescription. Rather, I hoped to deliver something else: a framework for understanding how habits work and a guide to experimenting with how they might change. Some habits yield easily to analysis and influence. Others are more complex and obstinate, and require prolonged study. And for others, change is a process that never fully concludes.

But that doesn't mean it can't occur. Each chapter in this book explains a different aspect of why habits exist and how they function. The framework described in this appendix is an attempt to distill, in a very basic way, the tactics that researchers have found for diagnosing and shaping habits within our own lives. This isn't meant to be comprehensive. This is merely a practical guide, a place to start. And paired with deeper lessons from this book's chapters, it's a manual for where to go next.

Change might not be fast and it isn't always easy. But with time and effort, almost any habit can be reshaped.

THE FRAMEWORK:
- Identify the routine
- Experiment with rewards
- Isolate the cue
- Have a plan

STEP ONE: IDENTIFY THE ROUTINE

The MIT researchers in chapter 1 discovered a simple neurological loop at the core of every habit, a loop that consists of three parts: a cue, a routine, and a reward.

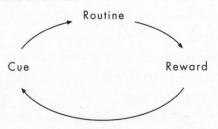

To understand your own habits, you need to identify the compo-nents of your loops. Once you have diagnosed the habit loop of a particular behavior, you can look for ways to supplant old vices with new routines.

As an example, let's say you have a bad habit, like I did when I started researching this book, of going to the cafeteria and buying a chocolate chip cookie every afternoon. Let's say this habit has caused you to gain a few pounds. In fact, let's say this habit has caused you to gain exactly eight pounds, and that your wife has made a few pointed comments. You've tried to force yourself to stop—you even went so far as to put a Post-it on your computer that reads NO MORE COOKIES.

But every afternoon you manage to ignore that note, get up, wan-der toward the cafeteria, buy a cookie, and, while chatting with col-leagues around the cash register, eat it. It feels good, and then it feels bad. Tomorrow, you promise yourself, you'll muster the will-power to resist. Tomorrow will be different.

But tomorrow the habit takes hold again.

How do you start diagnosing and then changing this behavior? By figuring out the habit loop. And the first step is to identify the routine. In this cookie scenario—as with most habits—the routine is the most obvious aspect: It's the behavior you want to change. Your routine is that you get up from your desk in the afternoon, walk to the cafeteria, buy a chocolate chip cookie, and eat it while chatting with friends. So that's what you put into the loop:

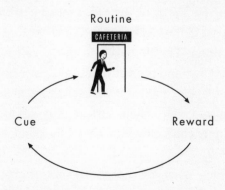

Next, some less obvious questions: What's the cue for this routine? Is it hunger? Boredom? Low blood sugar? That you need a break before plunging into another task?

And what's the reward? The cookie itself? The change of scenery? The temporary distraction? Socializing with colleagues? Or the burst of energy that comes from that blast of sugar?

To figure this out, you'll need to do a little experimentation.

STEP TWO: EXPERIMENT WITH REWARDS

Rewards are powerful because they satisfy cravings. But we're often not conscious of the cravings that drive our behaviors. When the Febreze marketing team discovered that consumers desired a fresh scent at the end of a cleaning ritual, for example, they had found a craving that no one even knew existed. It was hiding in plain sight. Most cravings are like this: obvious in retrospect, but incredibly hard to see when we are under their sway.

To figure out which cravings are driving particular habits, it's useful to experiment with different rewards. This might take a few days, or a week, or longer. During that period, you shouldn't feel any pressure to make a real change—think of yourself as a scientist in the data collection stage.

On the first day of your experiment, when you feel the urge to go to the cafeteria and buy a cookie, adjust your routine so it delivers a different reward. For instance, instead of walking to the cafeteria, go outside, walk around the block, and then go back to your desk without eating anything. The next day, go to the cafeteria and buy a donut, or a candy bar, and eat it at your desk. The next day, go to the cafeteria, buy an apple, and eat it while chatting with your friends. Then, try a cup of coffee. Then, instead of going to the cafeteria, walk over to your friend's office and gossip for a few minutes and go back to your desk.

You get the idea. What you choose to do *instead* of buying a cookie isn't important. The point is to test different hypotheses to determine which craving is driving your routine. Are you craving the cookie itself, or a break from work? If it's the cookie, is it because you're hungry? (In which case the apple should work just as well.) Or is it because you want the burst of energy the cookie provides? (And so the coffee should suffice.) Or are you wandering up to the cafeteria as an excuse to socialize, and the cookie is just a convenient excuse? (If so, walking to someone's desk and gossiping for a few minutes should satisfy the urge.)

As you test four or five different rewards, you can use an old trick to look for patterns: After each activity, jot down on a piece of paper the first three things that come to mind when you get back to your desk. They can be emotions, random thoughts, reflections on how you're feeling, or just the first three words that pop into your head.

RELAXED   SAW   NOT
            FLOWERS   HUNGRY

Then, set an alarm on your watch or computer for fifteen minutes. When it goes off, ask yourself: Do you still feel the urge for that cookie?

The reason why it's important to write down three things—even if they are meaningless words—is twofold. First, it forces a momentary awareness of what you are thinking or feeling. Just as Mandy, the nail biter in chapter 3, carried around a note card filled with hash marks to force her into awareness of her habitual urges, so writing three words forces a moment of attention. What's more, studies show that writing down a few words helps in later recalling what you were thinking at that moment. At the end of the experiment, when you review your notes, it will be much easier to remember what you were thinking and feeling at that precise instant, because your scribbled words will trigger a wave of recollection.

And why the fifteen-minute alarm? Because the point of these tests is to determine the reward you're craving. If, fifteen minutes after eating a donut, you *still* feel an urge to get up and go to the cafeteria, then your habit isn't motivated by a sugar craving. If, after gossiping at a colleague's desk, you still want a cookie, then the need for human contact isn't what's driving your behavior.

On the other hand, if fifteen minutes after chatting with a friend, you find it easy to get back to work, then you've identified the reward—temporary distraction and socialization—that your habit sought to satisfy.

By experimenting with different rewards, you can isolate what you are *actually* craving, which is essential in redesigning the habit.

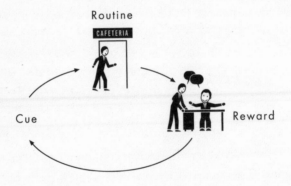

Once you've figured out the routine and the reward, what remains is identifying the cue.

STEP THREE: ISOLATE THE CUE

About a decade ago, a psychologist at the University of Western Ontario tried to answer a question that had bewildered social scientists for years: Why do some eyewitnesses of crimes misremember what they see, while other recall events accurately?

The recollections of eyewitnesses, of course, are incredibly important. And yet studies indicate that eyewitnesses often misremember what they observe. They insist that the thief was a man, for instance, when she was wearing a skirt; or that the crime occurred at dusk, even though police reports say it happened at 2:00 in the afternoon. Other eyewitnesses, on the other hand, can remember the crimes they've seen with near-perfect recall.

Dozens of studies have examined this phenomena, trying to determine why some people are better eyewitnesses than others. Researchers theorized that some people simply have better memories, or that a crime that occurs in a familiar place is easier to recall. But those theories didn't test out—people with strong and weak memories, or more and less familiarity with the scene of a crime, were equally liable to misremember what took place.

The psychologist at the University of Western Ontario took a different approach. She wondered if researchers were making a mistake by focusing on what questioners and witnesses had said, rather than *how* they were saying it. She suspected there were subtle cues that were influencing the questioning process. But when she watched videotape after videotape of witness interviews, looking for these cues, she couldn't see anything. There was so much activity in each interview—all the facial expressions, the different ways the questions were posed, the fluctuating emotions—that she couldn't detect any patterns.

So she came up with an idea: She made a list of a few elements she would focus on—the questioners' tone, the facial expressions of the witness, and how close the witness and the questioner were sitting to each other. Then she removed any information that would distract her from those elements. She turned down the volume on the television so instead of hearing words, all she could detect was the tone of the questioner's voice. She taped a sheet of paper over the questioner's face, so all she could see was the witnesses' expressions. She held a tape measure to the screen to measure their distance from each other.

And once she started studying these specific elements, patterns leapt out. She saw that witnesses who misremembered facts usually were questioned by cops who used a gentle, friendly tone. When witnesses smiled more, or sat closer to the person asking the questions, they were more likely to misremember.

In other words, when environmental cues said "we are friends"—a gentle tone, a smiling face—the witnesses were more likely to misremember what had occurred. Perhaps it was because, subconsciously, those friendship cues triggered a habit to please the questioner.

But the importance of this experiment is that those same tapes had been watched by dozens of other researchers. Lots of smart people had seen the same patterns, but no one had recognized them before. Because there was *too much* information in each tape to see a subtle cue.

Once the psychologist decided to focus on only three categories of behavior, however, and eliminate the extraneous information, the patterns leapt out.

Our lives are the same way. The reason why it is so hard to identify the cues that trigger our habits is because there is too much information bombarding us as our behaviors unfold. Ask yourself, do you eat breakfast at a certain time each day because you are hungry? Or because the clock says 7:30? Or because your kids have started eating? Or because you're dressed, and that's when the breakfast habit kicks in?

When you automatically turn your car left while driving to work, what triggers that behavior? A street sign? A particular tree? The knowledge that this is, in fact, the correct route? All of them together? When you're driving your kid to school and you find that you've absentmindedly started taking the route to work—rather than to the school—what caused the mistake? What was the cue that caused the "drive to work" habit to kick in, rather than the "drive to school" pattern?

To identify a cue amid the noise, we can use the same system as the psychologist: Identify categories of behaviors ahead of time to scrutinize in order to see patterns. Luckily, science offers some help in this regard. Experiments have shown that almost all habitual cues fit into one of five categories:

Location
Time
Emotional state
Other people
Immediately preceding action

So if you're trying to figure out the cue for the "going to the cafeteria and buying a chocolate chip cookie" habit, you write down five things the moment the urge hits (these are my actual notes from when I was trying to diagnose my habit):

Where are you? (sitting at my desk)
What time is it? (3:36 P.M.)
What's your emotional state? (bored)
Who else is around? (no one)
What action preceded the urge? (answered an email)

The next day:

Where are you? (walking back from the copier)
What time is it? (3:18 P.M.)
What's your emotional state? (happy)
Who else is around? (Jim from Sports)
What action preceded the urge? (made a photocopy)

The third day:

Where are you? (conference room)
What time is it? (3:41 P.M.)

What's your emotional state? (tired, excited about the project
I'm working on)
Who else is around? (editors who are coming to this meeting)
What action preceded the urge? (I sat down because the
meeting is about to start)

Three days in, it was pretty clear which cue was triggering my
cookie habit—I felt an urge to get a snack at a certain time of day. I
had already figured out, in step two, that it wasn't hunger driving my
behavior. The reward I was seeking was a temporary distraction—the
kind that comes from gossiping with a friend. And the habit, I now
knew, was triggered between 3:00 and 4:00.

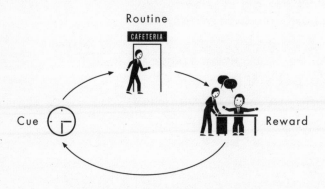

STEP FOUR: HAVE A PLAN

Once you've figured out your habit loop—you've identified the reward driving your behavior, the cue triggering it, and the routine
itself—you can begin to shift the behavior. You can change to a better routine by planning for the cue and choosing a behavior that
delivers the reward you are craving. What you need is a plan.

In the prologue, we learned that a habit is a choice that we deliberately make at some point, and then stop thinking about, but continue doing, often every day.

Put another way, a habit is a formula our brain automatically follows: When I see CUE, I will do ROUTINE in order to get a REWARD.

To re-engineer that formula, we need to begin making choices again. And the easiest way to do this, according to study after study, is to have a plan. Within psychology, these plans are known as "implementation intentions."

Take, for instance, my cookie-in-the-afternoon habit. By using this framework, I learned that my cue was roughly 3:30 in the afternoon. I knew that my routine was to go to the cafeteria, buy a cookie, and chat with friends. And, through experimentation, I had learned that it wasn't really the cookie I craved—rather, it was a moment of distraction and the opportunity to socialize.

So I wrote a plan:

> At 3:30, every day, I will walk to a friend's desk
> and talk for 10 minutes.

To make sure I remembered to do this, I set the alarm on my watch for 3:30.

It didn't work immediately. There were some days I was too busy and ignored the alarm, and then fell off the wagon. Other times it seemed like too much work to find a friend willing to chat—it was easier to get a cookie, and so I gave in to the urge. But on those days that I abided by my plan—when my alarm went off, I forced myself to walk to a friend's desk and chat for ten minutes—I found that I ended the workday feeling better. I hadn't gone to the cafeteria, I hadn't eat a cookie, and I felt fine. Eventually, it got be automatic: when the alarm rang, I found a friend and ended the day feeling a small, but real, sense of accomplishment. After a few weeks, I hardly thought about the routine anymore. And when I couldn't find anyone to chat with, I went to the cafeteria and bought tea and drank it with friends.

That all happened about six months ago. I don't have my watch anymore—I lost it at some point. But at about 3:30 every day, I absentmindedly stand up, look around the newsroom for someone to talk to, spend ten minutes gossiping about the news, and then go back to my desk. It occurs almost without me thinking about it. It has become a habit.

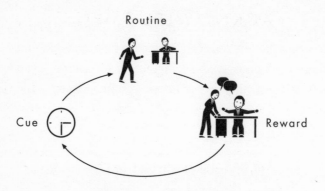

Obviously, changing some habits can be more difficult. But this framework is a place to start. Sometimes change takes a long time. Sometimes it requires repeated experiments and failures. But once you understand how a habit operates—once you diagnose the cue, the routine and the reward—you gain power over it.

ACKNOWLEDGMENTS

I have been undeservedly lucky throughout my life to work with people who are more talented than I am, and to get to steal their wisdom and gracefulness and pass it off as my own.

Which is why you are reading this book, and why I have so many people to thank.

Andy Ward acquired *The Power of Habit* before he even started as an editor at Random House. At the time, I did not know that he was a kind, generous, and amazingly—astoundingly—talented editor. I'd heard from some friends that he had elevated their prose and held their hands so gracefully they almost forgot the touch. But I figured they were exaggerating, since many of them were drinking at the time. Dear reader: it's all true. Andy's humility, patience and—most of all—the work he puts into being a good friend make everyone around him want to be a better person. This book is as much his as mine, and I am thankful that I had a chance to know, work with, and learn from him. Equally, I owe an enormous debt to some obscure deity for landing me at Random House under the wise guidance of Susan Kamil, the leadership of Gina Centrello, and the advice and efforts of Avideh Bashirrad, Tom Perry, Sanyu Dillon, Sally Marvin, Barbara Fillon, Maria Braeckel, Erika Greber, and the ever-patient Kaela Myers.

A similar twist of fortune allowed me to work with Scott Moyers, Andrew Wylie, and James Pullen at the Wylie Agency. Scott's counsel and friendship—as many writers know—is as invaluable as it is generous. Scott has moved back into the editorial world, and readers everywhere should consider themselves lucky. Andrew Wylie is always steadfast and astute in making the world safer (and more comfortable) for his writers, and I am enormously grateful. And James Pullen has helped me understand how to write in languages I didn't know existed.

Additionally, I owe an enormous amount to the New York Times. A huge thanks goes to Larry Ingrassia, *The Times*' business editor, whose friendship, advice and understanding allowed me to write this book, and to commit journalism among so many other talented reporters in an atmosphere where our work—and *The Times*' mission—is constantly elevated by his example. Vicki Ingrassia, too, has been a wonderful support. As any writer who has met Adam Bryant knows, he is an amazing advocate and friend, with gifted hands. And it is a privilege to work for Bill Keller, Jill Abramson, Dean Baquet and Glenn Kramon, and to follow their examples of how journalists should carry themselves through the world.

A few other thanks: I'm indebted to my *Times* colleagues Dean Murphy, Winnie O'Kelly, Jenny Anderson, Rick Berke, Andrew Ross Sorkin, David Leonhardt, Walt Bogdanich, David Gillen, Eduardo Porter, Jodi Kantor, Vera Titunik, Amy O'Leary, Peter Lattman, David Segal, Christine Haughney, Jenny Schussler, Joe Nocera and Jim Schacter (both of whom read chapters for me), Jeff Cane, Michael Barbaro and others who have been so generous with their friendship and their ideas.

Similarly, I'm thankful to Alex Blumberg, Adam Davidson, Paula Szuchman, Nivi Nord, Alex Berenson, Nazanin Rafsanjani, Brendan Koerner, Nicholas Thompson, Kate Kelly, Sarah Ellison, Kevin Bleyer, Amanda Schaffer, Dennis Potami, James Wynn, Noah Kotch, Greg Nelson, Caitlin Pike, Jonathan Klein, Amanda Klein, Donnan Steele, Stacey Steele, Wesley Morris, Adir Waldman, Rich Frankel, Jennifer Couzin, Aaron Bendikson, Richard Rampell, Mike Bor, David Lewicki, Beth Waltemath, Ellen Martin, Russ Uman, Erin Brown, Jeff Norton, Raj De Datta, Ruben Sigala, Dan Costello, Peter Blake, Peter Goodman, Alix Spiegel, Susan Dominus, Jenny Rosenstrach, Jason Woodard, Taylor Noguera, and Matthew Bird, who all provided support and guidance. The book's cover, and wonderful interior graphics, come from the mind of the incredibly talented Anton Ioukhnovets.

I also owe a debt to the many people who were generous with their time in reporting this book. Many are mentioned in the notes, but I wanted to give additional thanks to Tom Andrews at SYPartners, Tony Dungy and DJ Snell, Paul O'Neill, Warren Bennis, Rick Warren, Anne Krumm, Paco Underhill, Larry Squire, Wolfram Schultz, Ann Graybiel, Todd Heatherton, J. Scott Tonigan, Taylor Branch, Bob Bowman, Travis Leach, Howard Schultz, Mark Muraven, Angela Duckworth, Jane Bruno, Reza Habib, Patrick Mulkey and Terry Noffsinger. I was aided enormously by researchers and fact checkers, including Dax Proctor, Josh Friedman, Cole Louison, Alexander Provan and Neela Saldanha.

I am forever thankful to Bob Sipchen, who gave me my first real job in journalism, and am sorry that I won't be able to share this book with two friends lost too early, Brian Ching and L. K. Case.

Finally, my deepest thanks are to my family. Katy Duhigg, Jacquie Jenkusky, David Duhigg, Toni Martorelli, Daniel Duhigg, Alexandra Alter, and Jake Goldstein have been wonderful friends. My sons, Oliver and John Harry, have been sources of inspiration and sleeplessness. My parents, John and Doris, encouraged me from a young age to write, even as I was setting things on fire and giving them reason to figure that future correspondence might be on prison stationary.

And, of course, my wife, Liz, whose constant love, support, guidance, intelligence and friendship made this book possible.

—September, 2011

A NOTE ON SOURCES

The reporting in this book is based on hundreds of interviews, and thousands more papers and studies. Many of those sources are detailed in the text itself or the notes, along with guides to additional resources for interested readers.

In most situations, individuals who provided major sources of information or who published research that was integral to reporting were provided with an opportunity—after reporting was complete—to review facts and offer additional comments, address discrepancies, or register issues with how information is portrayed. Many of those comments are reproduced in the notes. (No source was given access to the book's complete text—all comments are based on summaries provided to sources.)

In a very small number of cases, confidentiality was extended to sources who, for a variety of reasons, could not speak on a for-attribution basis. In a very tiny number of instances, some identifying characteristics have been withheld or slightly modified to conform with patient privacy laws or for other reasons.

NOTES

xii **So they measured subjects' vital signs** Reporting for Lisa Allen's story is based on interviews with Allen. This research study is ongoing and unpublished, and thus researchers were not available for interviews. Basic outcomes, however, were confirmed by studies and interviews with scientists working on similar projects, including A. DelParigi et al., "Successful Dieters Have Increased Neural Activity in Cortical Areas Involved in the Control of Behavior," *International Journal of Obesity* 31 (2007): 440–48; Duc Son NT Le et al., "Less Activation in the Left Dorsolateral Prefrontal Cortex in the Reanalysis of the Response to a Meal in Obese than in Lean Women and Its Association with Successful Weight Loss," *American Journal of Clinical Nutrition* 86, no. 3 (2007): 573–79; A. DelParigi et al., "Persistence of Abnormal Neural Responses to a Meal in Postobese Individuals," *International Journal of Obesity* 28 (2004): 370–77; E. Stice et al., "Relation of Reward from Food Intake and Anticipated Food Intake to Obesity: A Functional Magnetic Resonance Imaging Study," *Journal of Abnormal Psychology* 117, no. 4 (November 2008): 924–35; A. C. Janes et al., "Brain fMRI Reactivity to Smoking-Related Images Before and During Extended Smoking Abstinence," *Experimental and Clinical Psychopharmacology* 17 (December 2009): 365–73; D. McBride et al., "Effects of Expectancy and Abstinence on the Neural Response to Smoking Cues in Cigarette Smokers: An fMRI Study," *Neuropsychopharmacology* 31 (December 2006): 2728–38; R. Sinha and C. S. Li, "Imaging Stress- and Cue-Induced Drug and Alcohol Craving: Association with Relapse and Clinical Implications," *Drug and Alcohol Review* 26, no. 1 (January 2007): 25–31; E. Tricomi, B. W. Balleine, and J. P. O'Doherty, "A Specific Role for Posterior Dorsolateral Striatum in Human Habit Learning," *European Journal of Neuroscience* 29, no. 11 (June 2009): 2225–32; D. Knoch, P. Bugger, and M. Regard, "Suppressing Versus Releasing a Habit: Frequency-Dependent Effects of Prefrontal Transcranial Magnetic Stimulation," *Cerebral Cortex* 15, no. 7 (July 2005): 885–87.

xv **"All our life, so far as it has"** William James, *Talks to Teachers on Psychology and to Students on Some of Life's Ideals,* originally published in 1899.

xvi **One paper published** Bas Verplanken and Wendy Wood, "Interventions to Break and Create Consumer Habits," *Journal of Public Policy and Marketing* 25, no. 1 (2006): 90–103; David T. Neal, Wendy Wood, and Jeffrey M. Quinn, "Habits—A Repeat Performance," *Current Directions in Psychological Science* 15, no. 4 (2006): 198–202.

xvii **The U.S. military, it occurred to me** For my understanding of the fascinating topic of the military's use of habit training, I am indebted to Dr. Peter Schifferle at the School of Advanced Military Studies, Dr. James Lussier, and the many commanders and soldiers who were generous with their time both in Iraq and at SAMS. For more on this topic, see Scott B. Shadrick and James W. Lussier, "Assessment of the Think Like a Commander Training Program," U.S. Army Research Institute for the Behavioral and Social Sciences Research Report 1824, July 2004; Scott B. Shadrick et al., "Positive Transfer of Adaptive Battlefield Thinking Skills," U.S. Army Research Institute for the Behavioral and Social Sciences Research Report 1873, July 2007; Thomas J. Carnahan et al., "Novice Versus Expert Command Groups: Preliminary Findings and Training Implications for Future Combat Systems," U.S. Army Research Institute for the Behavioral and Social Sciences Research Report 1821, March 2004; Carl W. Lickteig et al., "Human Performance Essential to Battle Command: Report on Four Future Combat Systems Command and Control Experiments," U.S. Army Research Institute for the Behavioral and Social Sciences Research Report 1812, November 2003; and Army Field Manual 5–2 20, February 2009.

CHAPTER ONE

3 **six feet tall** Lisa Stefanacci et al., "Profound Amnesia After Damage to the Medial Temporal Lobe: A Neuroanatomical and Neuropsychological Profile of Patient E.P.," *Journal of Neuroscience* 20, no. 18 (2000): 7024–36.

3 **"Who's Michael?"** I am indebted to the Pauly and Rayes families, as well as the Squire laboratory and coverage such as Joshua Foer, "Remember This,"

National Geographic, November 2007, 32–57; "Don't Forget," *Scientific American Frontiers*, television program, produced by Chedd-Angier Production Company, PBS, episode first aired May 11, 2004, hosted by Alan Alda; "Solved: Two Controversial Brain Teasers," *Bioworld Today*, August 1999; David E. Graham, "UCSD Scientist Unlocks Working of Human Memory," *The San Diego Union-Tribune*, August 12, 1999.

4 **The sample from Eugene's spine** Richard J. Whitley and David W. Kimberlan, "Viral Encephalitis," *Pediatrics in Review* 20, no. 6 (1999): 192–98.

7 **was seven years old** Some published papers say that H.M. was injured at age nine; others say seven.

7 **he was hit by a bicycle** Previous research indicates that H.M. was hit by a bicycle. New documents, as yet unpublished, indicate he may have fallen off a bike.

7 **landed hard on his head** Luke Dittrich, "The Brain That Changed Everything," *Esquire*, October 2010.

7 **He was smart** Eric Hargreaves, "H.M.," *Page O'Neuroplasticity*, http://homepages.nyu.edu/~eh597/HM.htm.

7 **When the doctor proposed cutting** Benedict Carey, "H. M., Whose Loss of Memory Made Him Unforgettable, Dies," *The New York Times*, December 5, 2008.

7 **with a small straw** This was a common practice at the time.

8 **He introduced himself to his doctors** Dittrich, "The Brain That Changed Everything"; Larry R. Squire, "Memory and Brain Systems: 1969–2009," *Journal of Neuroscience* 29, no. 41 (2009): 12711–26; Larry R. Squire, "The Legacy of Patient H.M. for Neuroscience," *Neuron* 61, no. 1 (2009): 6–9.

10 **transformed our understanding of habits' power** Jonathan M. Reed et al., "Learning About Categories That Are Defined by Object-Like Stimuli Despite Impaired Declarative Memory," *Behavioral Neuroscience* 113 (1999): 411–19; B. J. Knowlton, J. A. Mangels, and L. R. Squire, "A Neostriatal Habit Learning System in Humans," *Science* 273 (1996): 1399–1402; P. J. Bayley, J. C. Frascino, and L. R. Squire, "Robust Habit Learning in the Absence of Awareness and Independent of the Medial Temporal Lobe," *Nature* 436 (2005): 550–53.

13 **a golf ball–sized** B. Bendriem et al., "Quantitation of the Human Basal Ganglia with Positron Emission Tomography: A Phantom Study of the Effect of Contrast and Axial Positioning," *IEEE Transactions on Medical Imaging* 10, no. 2 (1991): 216–22.

14 **an oval of cells** G. E. Alexander and M. D. Crutcher, "Functional Architecture of Basal Ganglia Circuits: Neural Substrates of Parallel Processing," *Trends in Neurosciences* 13 (1990): 266–71; André Parent and Lili-Naz Hazrati, "Func-

tional Anatomy of the Basal Ganglia," *Brain Research Reviews* 20 (1995): 91–127; Roger L. Albin, Anne B. Young, and John B. Penney, "The Functional Anatomy of Basal Ganglia Disorders," *Trends in Neurosciences* 12 (1989): 366–75.

14   diseases such as Parkinson's Alain Dagher and T. W. Robbins, "Personality, Addiction, Dopamine: Insights from Parkinson's Disease," *Neuron* 61 (2009): 502–10.

14   to open food containers I am indebted to the following sources for expanding my understanding of the work at the MIT labs, the basal ganglia, and its role in habits and memory: F. Gregory Ashby and John M. Ennis, "The Role of the Basal Ganglia in Category Learning," *Psychology of Learning and Motivation* 46 (2006): 1–36; F. G. Ashby, B. O. Turner, and J. C. Horvitz, "Cortical and Basal Ganglia Contributions to Habit Learning and Automaticity," *Trends in Cognitive Sciences* 14 (2010): 208–15; C. Da Cunha and M. G. Packard, "Preface: Special Issue on the Role of the Basal Ganglia in Learning and Memory," *Behavioural Brain Research* 199 (2009): 1–2; C. Da Cunha et al., "Learning Processing in the Basal Ganglia: A Mosaic of Broken Mirrors," *Behavioural Brain Research* 199 (2009): 157–70; M. Desmurget and R. S. Turner, "Motor Sequences and the Basal Ganglia: Kinematics, Not Habits," *Journal of Neuroscience* 30 (2010): 7685–90; J. J. Ebbers and N. M. Wijnberg, "Organizational Memory: From Expectations Memory to Procedural Memory," *British Journal of Management* 20 (2009): 478–90; J. A. Grahn, J. A. Parkinson, and A. M. Owen, "The Role of the Basal Ganglia in Learning and Memory: Neuropsychological Studies," *Behavioural Brain Research* 199 (2009): 53–60; Ann M. Graybiel, "The Basal Ganglia: Learning New Tricks and Loving It," *Current Opinion in Neurobiology* 15 (2005): 638–44; Ann M. Graybiel, "The Basal Ganglia and Chunking of Action Repertoires," *Neurobiology of Learning and Memory* 70, nos. 1–2 (1998): 119–36; F. Gregory Ashby and V. Valentin, "Multiple Systems of Perceptual Category Learning: Theory and Cognitive Tests," in *Handbook of Categorization in Cognitive Science*, ed. Henri Cohen and Claire Lefebvre (Oxford: Elsevier Science, 2005); S. N Haber and M. Johnson Gdowski, "The Basal Ganglia," in *The Human Nervous System*, 2nd ed., ed. George Paxinos and Jürgen K. Mai (San Diego: Academic Press, 2004), 676–738; T. D. Barnes et al., "Activity of Striatal Neurons Reflects Dynamic Encoding and Recoding of Procedural Memories," *Nature* 437 (2005): 1158–61; M. Laubach, "Who's on First? What's on Second? The Time Course of Learning in Corticostriatal Systems," *Trends in Neurosciences* 28 (2005): 509–11; E. K. Miller and T. J. Buschman, "Bootstrapping Your Brain: How Interactions Between the Frontal Cortex and Basal Ganglia May Produce Organized Actions and Lofty Thoughts," in *Neurobiology of Learning and Memory*, 2nd ed., ed. Raymond P. Kesner and Joe L. Martinez (Burlington, Vt.: Academic Press, 2007), 339–54; M. G. Packard, "Role of Basal Ganglia in Habit Learning and Memory: Rats, Monkeys, and

Humans," in *Handbook of Behavioral Neuroscience*, ed. Heinz Steiner and Kuei Y. Tseng, 561–69; D. P. Salmon and N. Butters, "Neurobiology of Skill and Habit Learning," *Current Opinion in Neurobiology* 5 (1995): 184–90; D. Shohamy et al., "Role of the Basal Ganglia in Category Learning: How Do Patients with Parkinson's Disease Learn?" *Behavioral Neuroscience* 118 (2004): 676–86; M. T. Ullman, "Is Broca's Area Part of a Basal Ganglia Thalamocortical Circuit?" *Cortex* 42 (2006): 480–85; N. M. White, "Mnemonic Functions of the Basal Ganglia," *Current Opinion in Neurobiology* 7 (1997): 164–69.

14 The maze was structured Ann M. Graybiel, "Overview at Habits, Rituals, and the Evaluative Brain," *Annual Review of Neuroscience* 31 (2008): 359–87; T. D. Barnes et al., "Activity of Striatal Neurons Reflects Dynamic Encoding and Recoding of Procedural Memories," *Nature* 437 (2005): 1158–61; Ann M. Graybiel, "Network-Level Neuroplasticity in Cortico-Basal Ganglia Pathways," *Parkinsonism and Related Disorders* 10 (2004): 293–96; N. Fujii and Ann M. Graybiel, "Time-Varying Covariance of Neural Activities Recorded in Striatum and Frontal Cortex as Monkeys Perform Sequential-Saccade Tasks," *Proceedings of the National Academy of Sciences* 102 (2005): 9032–37.

16 To see this capacity in action The graphs in this chapter have been simplified to exhibit salient aspects. However, a full description of these studies can be found among Dr. Graybiel's papers and lectures.

17 root of how habits form Ann M. Graybiel, "The Basal Ganglia and Chunking of Action Repertoires," *Neurobiology of Learning and Memory* 70 (1998): 119–36.

19 a habit is born For more, see A. David Smith and J. Paul Bolam, "The Neural Network of the Basal Ganglia as Revealed by the Study of Synaptic Connections of Identified Neurones," *Trends in Neurosciences* 13 (1990): 259–65; John G. McHaffie et al., "Subcortical Loops Through the Basal Ganglia," *Trends in Neurosciences* 28 (2005): 401–7; Ann M. Graybiel, "Neurotransmitters and Neuromodulators in the Basal Ganglia," *Trends in Neurosciences* 13 (1990): 244–54; J. Yelnik, "Functional Anatomy of the Basal Ganglia," *Movement Disorders* 17 (2002): 15–21.

20 The problem is that your brain For more, see Catherine A. Thorn et al., "Differential Dynamics of Activity Changes in Dorsolateral and Dorsomedial Striatal Loops During Learning," *Neuron* 66 (2010): 781–95; Ann M. Graybiel, "The Basal Ganglia: Learning New Tricks and Loving It," *Current Opinion in Neurobiology* 15 (2005): 638–44.

22 In each pairing, one piece For more, see Peter J. Bayley, Jennifer C. Frascino, and Larry R. Squire, "Robust Habit Learning in the Absence of Awareness and Independent of the Medial Temporal Lobe," *Nature* 436 (2005): 550–53; J. M. Reed et al., "Learning About Categories That Are Defined by Object-Like

Stimuli Despite Impaired Declarative Memory," *Behavioral Neuroscience* 133 (1999): 411–19; B. J. Knowlton, J. A. Mangels, and L. R. Squire, "A Neostriatal Habit Learning System in Humans," *Science* 273 (1996): 1399–1402.

24 Squire's experiments with Eugene It is worth noting that Squire's work with Pauly is not limited to habits and has also provided insights into subjects such as spatial memory and the effects of priming on the brain. For a more complete discussion of discoveries made possible by Pauly, see Squire's home page at http://psychiatry.ucsd.edu/faculty/lsquire.html.

26 The habit was so ingrained For discussion, see Monica R. F. Hilario et al., "Endocannabinoid Signaling Is Critical for Habit Formation," *Frontiers in Integrative Neuroscience* 1 (2007): 6; Monica R. F. Hilario and Rui M. Costa, "High on Habits," *Frontiers in Neuroscience* 2 (2008): 208–17; A. Dickinson, "Appetitive-Aversive Interactions: Superconditioning of Fear by an Appetitive CS," *Quarterly Journal of Experimental Psychology* 29 (1977): 71–83; J. Lamarre and P. C. Holland, "Transfer of Inhibition After Serial Feature Negative Discrimination Training," *Learning and Motivation* 18 (1987): 319–42; P. C. Holland, "Differential Effects of Reinforcement of an Inhibitory Feature After Serial and Simultaneous Feature Negative Discrimination Training," *Journal of Experimental Psychology: Animal Behavior Processes* 10 (1984): 461–75.

26 When researchers at the University of North Texas Jennifer L. Harris, Marlene B. Schwartz, and Kelly D. Brownell, "Evaluating Fast Food Nutrition and Marketing to Youth," Yale Rudd Center for Food Policy and Obesity, 2010; H. Qin and V. R. Prybutok, "Determinants of Customer-Perceived Service Quality in Fast-Food Restaurants and Their Relationship to Customer Satisfaction and Behavioral Intentions," *The Quality Management Journal* 15 (2008): 35; H. Qin and V. R. Prybutok, "Service Quality, Customer Satisfaction, and Behavioral Intentions in Fast-Food Restaurants," *International Journal of Quality and Service Sciences* 1 (2009): 78. For more on this topic, see K. C. Berridge, "Brain Reward Systems for Food Incentives and Hedonics in Normal Appetite and Eating Disorders," in *Appetite and Body Weight*, ed. Tim C. Kirkham and Steven J. Cooper (Burlington, Vt.: Academic Press, 2007), 91–215; K. C. Berridge et al., "The Tempted Brain Eats: Pleasure and Desire Circuits in Obesity and Eating Disorders," *Brain Research* 1350 (2010): 43–64; J. M. Dave et al., "Relationship of Attitudes Toward Fast Food and Frequency of Fast-Food Intake in Adults," *Obesity* 17 (2009): 1164–70; S. A. French et al., "Fast Food Restaurant Use Among Adolescents: Associations with Nutrient Intake, Food Choices and Behavioral and Psychosocial Variables," *International Journal of Obesity and Related Metabolic Disorders* 25 (2001): 1823; N. Ressler, "Rewards and Punishments, Goal-Directed Behavior and Consciousness," *Neuroscience and Biobehavioral Reviews* 28 (2004): 27–39; T. J. Richards, "Fast Food, Addiction, and Market Power," *Journal of Agricultural and Resource*

Economics 32 (2007): 425–47; M. M. Torregrossa, J. J. Quinn, and J. R. Taylor, "Impulsivity, Compulsivity, and Habit: The Role of Orbitofrontal Cortex Revisited," *Biological Psychiatry* 63 (2008): 253–55; L. R. Vartanian, C. P. Herman, and B. Wansink, "Are We Aware of the External Factors That Influence Our Food Intake?" *Health Psychology* 27 (2008): 533–38; T. Yamamoto and T. Shimura, "Roles of Taste in Feeding and Reward," in *The Senses: A Comprehensive Reference*, ed. Allan I. Basbaum et al. (New York: Academic Press, 2008), 437–58; F. G. Ashby, B. O. Turner, and J. C. Horvitz, "Cortical and Basal Ganglia Contributions to Habit Learning and Automaticity," *Trends in Cognitive Sciences* 14 (2010): 208–15.

27 **All the better for tightening** K. C. Berridge and T. E. Robinson, "Parsing Reward," *Trends in Neurosciences* 26 (2003): 507–13; Kelly D. Brownell and Katherine Battle Horgen, *Food Fight: The Inside Story of the Food Industry, America's Obesity Crisis, and What We Can Do About It* (Chicago: Contemporary Books, 2004); Karl Weber, ed., *Food, Inc.: How Industrial Food Is Making Us Sicker, Fatter, and Poorer—and What You Can Do About It* (New York: Public Affairs, 2004); Ronald D. Michman and Edward M. Mazze, *The Food Industry Wars: Marketing Triumphs and Blunders* (Westport, Conn.: Quorum Books, 1998); M. Nestle, *Food Politics: How the Food Industry Influences Nutrition and Health* (Berkeley: University of California Press, 2002); D. R. Reed and A. Knaapila, "Genetics of Taste and Smell: Poisons and Pleasures," in *Progress in Molecular Biology and Translational Science*, ed. Claude Bouchard (New York: Academic Press); N. Ressler, "Rewards and Punishments, Goal-Directed Behavior and Consciousness," *Neuroscience and Biobehavioral Reviews* 28 (2004): 27–39; T. Yamamoto and T. Shimura, "Roles of Taste in Feeding and Reward," in *The Senses: A Comprehensive Reference*, ed. Allan I. Basbaum et al. (New York: Academic Press, 2008), 437–58.

CHAPTER TWO

31 **Hopkins would consent to** For the history of Hopkins, Pepsodent, and dental care in the United States, I am indebted to Scott Swank, curator at the Dr. Samuel D. Harris National Museum of Dentistry; James L. Gutmann, DDS; and David A. Chemin, editor of the *Journal of the History of Dentistry*. In addition, I drew heavily on James Twitchell, *Twenty Ads That Shook the World* (New York: Three Rivers Press, 2000); the Dr. Samuel D. Harris National Museum of Dentistry; the *Journal of the History of Dentistry*; Mark E. Parry, "Crest Toothpaste: The Innovation Challenge," *Social Science Research Network*, October 2008; Robert Aunger, "Tooth Brushing as Routine Behavior," *International Dental Journal* 57 (2007): 364–76; Jean-Paul Claessen et al., "Designing Interventions to Improve Tooth Brushing," *International Dental Journal* 58 (2008):

307–20; Peter Miskell, "Cavity Protection or Cosmetic Perfection: Innovation and Marketing of Toothpaste Brands in the United States and Western Europe, 1955–1985," *Business History Review* 78 (2004): 29–60; James L. Gutmann, "The Evolution of America's Scientific Advancements in Dentistry in the Past 150 Years," *The Journal of the American Dental Association* 140 (2009): 8S–15S; Domenick T. Zero et al., "The Biology, Prevention, Diagnosis and Treatment of Dental Caries: Scientific Advances in the United States," *The Journal of the American Dental Association* 140 (2009): 25S–34S; Alyssa Picard, *Making of the American Mouth: Dentists and Public Health in the Twentieth Century* (New Brunswick, N.J.: Rutgers University Press, 2009); S. Fischman, "The History of Oral Hygiene Products: How Far Have We Come in 6,000 Years?" *Periodontology 2000* 15 (1997): 7–14; Vincent Vinikas, *Soft Soap, Hard Sell: American Hygiene in the Age of Advertisement* (Ames: University of Iowa Press, 1992).

32 As the nation had become wealthier H. A. Levenstein, *Revolution at the Table: The Transformation of the American Diet* (New York: Oxford University Press, 1988); Scott Swank, *Paradox of Plenty: The Social History of Eating in Modern America* (Berkeley: University of California Press, 2003).

32 hardly anyone brushed their teeth Alyssa Picard, *Making of the American Mouth: Dentists and Public Health in the Twentieth Century* (New Brunswick, N.J.: Rutgers University Press, 2009).

33 everyone from Shirley Temple For more on celebrity advertising of toothpaste, see Steve Craig, "The More They Listen, the More They Buy: Radio and the Modernizing of Rural America, 1930–1939," *Agricultural History* 80 (2006): 1–16.

33 By 1930, Pepsodent was sold Kerry Seagrave, *America Brushes Up: The Use and Marketing of Toothpaste and Toothbrushes in the Twentieth Century* (Jefferson, N.C.: McFarland, 2010); Alys Eve Weinbaum, et al., *The Modern Girl Around the World: Consumption, Modernity, and Globalization* (Durham, N.C.: Duke University Press, 2008), 28–30.

33 A decade after the first Scripps-Howard, *Market Records, from a Home Inventory Study of Buying Habits and Brand Preferences of Consumers in Sixteen Cities* (New York: Scripps-Howard Newspapers, 1938).

34 The film is a naturally occurring membrane C. McGaughey and E. C. Stowell, "The Adsorption of Human Salivary Proteins and Porcine Submaxillary Mucin by Hydroxyapatite," *Archives of Oral Biology* 12, no. 7 (1967): 815–28; Won-Kyu Park et al., "Influences of Animal Mucins on Lysozyme Activity in Solution and on Hydroxyapatite Surface," *Archives of Oral Biology* 51, no. 10 (2006): 861–69.

34 particularly Pepsodent—were worthless William J. Gies, "Experimental Studies of the Validity of Advertised Claims for Products of Public Importance in Relation to Oral Hygiene or Dental Therapeutics," *Journal of Dental Research* 2 (September 1920): 511–29.

35   Pepsodent removes the film! I am indebted to the Duke University digital collection of advertisements.

35   Pepsodent was one of the top-selling Kerry Seagrave, *America Brushes Up: The Use and Marketing of Toothpaste and Toothbrushes in the Twentieth Century* (Jefferson, N.C.: McFarland, 2010); Jeffrey L. Cruikshank and Arthur W. Schultz, *The Man Who Sold America: The Amazing (but True!) Story of Albert D. Lasker and the Creation of the Advertising Century* (Cambridge, Mass.: Harvard Business Press, 2010), 268–81.

35   best-selling toothpaste for more than Pepsodent was eventually outsold by Crest, which featured fluoride—the first ingredient in toothpaste that actually made it effective at fighting cavities.

35   A decade after Hopkins's ad campaign Peter Miskell, "Cavity Protection or Cosmetic Perfection: Innovation and Marketing of Toothpaste Brands in the United States and Western Europe, 1955–1985," *Business History Review* 78 (2004): 29–60.

36   Studies of people who have successfully H. Aarts, T. Paulussen, and H. Schaalma, "Physical Exercise Habit: On the Conceptualization and Formation of Habitual Health Behaviours," *Health Education Research* 3 (1997): 363–74.

36   Research on dieting says Krystina A. Finlay, David Trafimow, and Aimee Villarreal, "Predicting Exercise and Health Behavioral Intentions: Attitudes, Subjective Norms, and Other Behavioral Determinants," *Journal of Applied Social Psychology* 32 (2002): 342–56.

37   In the clothes-washing market alone Tara Parker-Pope, "P&G Targets Textiles Tide Can't Clean," *The Wall Street Journal*, April 29, 1998.

37   Its revenues topped $35 billion Peter Sander and John Slatter, *The 100 Best Stocks You Can Buy* (Avon, Mass.: Adams Business, 2009), 294.

39   They decided to call it Febreze The history of Febreze comes from interviews and articles, including "Procter & Gamble—Jager's Gamble," *The Economist*, October 28, 1999; Christine Bittar, "P&G's Monumental Repackaging Project," *Brandweek*, March 2000, 40–52; Jack Neff, "Does P&G Still Matter?" *Advertising Age* 71 (2000): 48–56; Roderick E. White and Ken Mark, "Procter & Gamble Canada: The Febreze Decision," Ivey School of Business, London, Ontario, 2001. Procter & Gamble was asked to comment on the reporting contained in this chapter, and in a statement said: "P&G is committed to ensuring the confidentiality of information shared with us by our consumers. Therefore, we are unable to confirm or correct information that you have received from sources outside of P&G."

41 **The second ad featured a woman** Christine Bittar, "Freshbreeze at P&G," *Brandweek*, October 1999.

41 **The cue: pet smells** American Veterinary Medical Association, market research statistics for 2001.

42 **So a new group of researchers joined** A. J. Lafley and Ram Charan, *The Game Changer: How You Can Drive Revenue and Profit Growth with Innovation* (New York: Crown Business, 2008).

44 **Rather than rats, however** An overview of Wolfram Schultz's research can be found in "Behavioral Theories and the Neurophysiology of Reward," *Annual Review of Psychology* 57 (2006): 87–115; Wolfram Schultz, Peter Dayan, and P. Read Montague, "A Neural Substrate of Prediction and Reward," *Science* 275 (1997): 1593–99; Wolfram Schultz, "Predictive Reward Signal of Dopamine Neurons," *Journal of Neurophysiology* 80 (1998): 1–27; L. Tremblya and Wolfram Schultz, "Relative Reward Preference in Primate Orbitofrontal Cortex," *Nature* 398 (1999): 704–8; Wolfram Schultz, "Getting Formal with Dopamine and Reward," *Neuron* 36 (2002): 241–63; W. Schultz, P. Apicella, and T. Ljungberg, "Responses of Monkey Dopamine Neurons to Reward and Conditioned Stimuli During Successive Steps of Learning a Delayed Response Task," *Journal of Neuroscience* 13 (1993): 900–913.

45 **he was experiencing happiness** It is important to note that Schultz does not claim that these spikes represent happiness. To a scientist, a spike in neural activity is just a spike, and assigning it subjective attributes is beyond the realm of provable results. In a fact-checking email, Schultz clarified: "We cannot talk about pleasure and happiness, as we don't know the feelings of an animal. . . . We try to avoid unsubstantiated claims and simply look at the facts." That said, as anyone who has ever seen a monkey, or a three-year-old human, receive some juice can attest, the result looks a lot like happiness.

47 **The anticipation and sense of craving** Schultz, in a fact-checking email, clarifies that his research focused not only on habits but on other behaviors as well: "Our data are not restricted to habits, which are one particular form of behavior. Rewards, and reward prediction errors, play a general role in all behaviors. Irrespective of habit or not, when we don't get what we expect, we feel disappointed. That we call a negative prediction error (the negative difference between what we get and what we expected)."

48 **Most food sellers locate their kiosks** Brian Wansink, *Mindless Eating: Why We Eat More Than We Think* (New York: Bantam, 2006); Sheila Sasser and David Moore, "Aroma-Driven Craving and Consumer Consumption Impulses," presentation, session 2.4, American Marketing Association Summer Educator Conference, San Diego, California, August 8–11, 2008; David Fields, "In Sales, Nothing You Say Matters," Ascendant Consulting, 2005.

48 **The habit loop is spinning because** Harold E. Doweiko, *Concepts of Chemical Dependency* (Belmont, Calif.: Brooks Cole, 2008), 362–82.

49 **how new habits are created** K. C. Berridge and M. L. Kringelbach, "Affective Neuroscience of Pleasure: Reward in Humans and Animals," *Psychopharmacology* 199 (2008): 457–80; Wolfram Schultz, "Behavioral Theories and the Neurophysiology of Reward," *Annual Review of Psychology* 57 (2006): 87–115.

50 **"wanting evolves into obsessive craving"** T. E. Robinson and K. C. Berridge, "The Neural Basis of Drug Craving: An Incentive-Sensitization Theory of Addiction," *Brain Research Reviews* 18 (1993): 247–91.

51 **In 2002 researchers at New Mexico** Krystina A. Finlay, David Trafimow, and Aimee Villarreal, "Predicting Exercise and Health Behavioral Intentions: Attitudes, Subjective Norms, and Other Behavioral Determinants," *Journal of Applied Social Psychology* 32 (2002): 342–56.

51 **The cue, in addition to triggering** Henk Aarts, Theo Paulussen, and Herman Schaalma, "Physical Exercise Habit: On the Conceptualization and Formation of Habitual Health Behaviours," *Health Education Research* 12 (1997): 363–74.

55 **Within a year, customers had spent** Christine Bittar, "Freshbreeze at P&G," *Brandweek*, October 1999.

57 **Unlike other pastes** Patent 1,619,067, assigned to Rudolph A. Kuever.

58 **Want to craft a new eating** J. Brug, E. de Vet, J. de Nooijer, and B. Verplanken, "Predicting Fruit Consumption: Cognitions, Intention, and Habits," *Journal of Nutrition Education and Behavior* 38 (2006): 73–81.

59 **The craving drove the habit** For a full inventory of studies from the National Weight Control Registry, see http://www.nwcr.ws/Research/published%20research.htm.

59 **Yet, while everyone brushes** D. I. McLean and R. Gallagher, "Sunscreens: Use and Misuse," *Dermatologic Clinics* 16 (1998): 219–26.

CHAPTER THREE

60 **The game clock at the far end** I am indebted to the time and writings of Tony Dungy and Nathan Whitacker, including *Quiet Strength: The Principles, Practices, and Priorities of a Winning Life* (Carol Stream, Ill.: Tyndale House, 2008); *The Mentor Leader: Secrets to Building People and Teams That Win Consistently* (Carol Stream, Ill.: Tyndale House, 2010); *Uncommon: Finding Your Path to Significance* (Carol Stream, Ill.: Tyndale House, 2011). I also owe a debt to Jene Bramel of Footballguys.com; Matthew Bowen of National Football Post and the St. Louis Rams, Green Bay Packers, Washington Redskins, and Buffalo Bills; Tim Layden of *Sports Illustrated* and his book *Blood, Sweat, and Chalk: The Ultimate Football Playbook: How the Great Coaches Built Today's Teams* (New York: Sports Illustrated, 2010); Pat Kirwan, *Take Your Eye Off the Ball: How to Watch Football by Knowing Where to Look* (Chicago: Triumph Books, 2010); Nunyo Demasio, "The Quiet Leader," *Sports Illustrated*, February 2007; Bill Plaschke, "Color Him Orange," *Los Angeles Times*, September 1, 1996; Chris Harry, "'Pups' Get to Bark for the Bucs," *Orlando Sentinel*, September 5, 2001; Jeff Legwold, "Coaches Find Defense in Demand," *Rocky Mountain News*, November 11, 2005; and Martin Fennelly, "Quiet Man Takes Charge with Bucs," *The Tampa Tribune*, August 9, 1996.

60   It's late on a Sunday I am indebted to Fox Sports for providing game tapes, and to Kevin Kernan, "The Bucks Stomp Here," *The San Diego Union-Tribune*, November 18, 1996; Jim Trotter, "Harper Says He's Done for Season," *The San Diego Union-Tribune*, November 18, 1996; Les East, "Still Worth the Wait," *The Advocate* (Baton Rouge, La.), November 21, 1996.

60   described as putting the "less" in "hopeless" Mitch Albom, "The Courage of Detroit," *Sports Illustrated*, September 22, 2009.

61   "America's Orange Doormat" Pat Yasinskas, "Behind the Scenes," *The Tampa Tribune*, November 19, 1996.

62   He knew from experience In a fact-checking letter, Dungy emphasized that these were not new strategies, but instead were approaches "I had learned from working with the Steelers in the seventies and eighties. What was unique, and what I think spread, was the idea of how to get those ideas across. . . . [My plan was] not overwhelming opponents with strategy or abundance of plays and formations but winning with execution. Being very sure of what we were doing and doing it well. Minimize the mistakes we would make. Playing with speed because we were not focusing on too many things."

64   When his strategy works For more on the Tampa 2 defense, see Rick Gosselin, "The Evolution of the Cover Two," *The Dallas Morning News*, November 3, 2005; Mohammed Alo, "Tampa 2 Defense," *The Football Times*, July 4, 2006; Chris Harry, "Duck and Cover," *Orlando Sentinel*, August 26, 2005; Jason Wilde, "What to Do with Tampa-2?" *Wisconsin State Journal*, September 22, 2005; Jim Thomas, "Rams Take a Run at Tampa 2," *St. Louis Post-Dispatch*, October 16, 2005; Alan Schmadtke, "Dungy's 'D' No Secret," *Orlando Sentinel*, September 6, 2006; Jene Bramel, "Guide to NFL Defenses," *The Fifth Down* (blog), *The New York Times*, September 6, 2010.

66   Sitting in the basement William L. White, *Slaying the Dragon* (Bloomington, Ill.: Lighthouse Training Institute, 1998).

66   named Bill Wilson Alcoholics Anonymous World Service, *The A.A. Service Manual Combined with Twelve Concepts for World Service* (New York: Alcoholics Anonymous, 2005); Alcoholics Anonymous World Service, *Alcoholics Anonymous: The Story of How Many Thousands of Men and Women Have Recovered from Alcoholism* (New York: Alcoholics Anonymous, 2001); Alcoholics Anonymous World Service, *Alcoholics Anonymous Comes of Age: A Brief History of A.A.* (New York: Alcoholics Anonymous, 1957); Alcoholics Anonymous World Service, *As Bill Sees It* (New York: Alcoholics Anonymous, 1967); Bill W., *Bill W.: My First 40 Years—An Autobiography by the Cofounder of Alcoholics Anonymous* (Hazelden Center City, Minn.: Hazelden Publishing, 2000); Francis Hartigan, *Bill W.: A Biography of Alcoholics Anonymous Cofounder Bill Wilson* (New York: Thomas Dunne Books, 2009).

67   He took a sip and felt Susan Cheever, *My Name Is Bill: Bill Wilson—His Life and the Creation of Alcoholics Anonymous* (New York: Simon and Schuster, 2004).

67   Wilson invited him over Ibid.

68   At that moment, he later wrote Ernest Kurtz, *Not-God: A History of Alcoholics Anonymous* (Hazelden Center City, Minn.: Hazelden Publishing, 1991).

68   An estimated 2.1 million people Data provided by AA General Service Office Staff, based on 2009 figures.

68   as many as 10 million alcoholics Getting firm figures on AA's membership or those who have achieved sobriety through the program is notoriously difficult, in part because membership is anonymous and in part because there is no requirement to register with a central authority. However, the 10 million person figure, based on conversations with AA researchers, seems reasonable (if unverifiable) given the program's long history.

68   What's interesting about AA In psychology, this kind of treatment—targeting habits—is often referred to under the umbrella term of "cognitive behavioral therapy," or in an earlier era, "relapse prevention." CBT, as it is generally used within the treatment community, often incorporates five basic techniques: (1) Learning, in which the therapist explains the illness to the patient and teaches the patient to identify the symptoms; (2) Monitoring, in which the patient uses a diary to monitor the behavior and the situations triggering it; (3) Competing response, in which the patient cultivates new routines, such as relaxation methods, to offset the problematic behavior; (4) Rethinking, in which a therapist guides the patient to reevaluate how the patient sees situations; and (5) Exposing, in which the therapist helps the patient expose him- or herself to situations that trigger the behavior.

69   What AA provides instead Writing about AA is always a difficult proposition, because the program has so many critics and supporters, and there are dozens of interpretations for how and why the program works. In an email, for instance, Lee Ann Kaskutas, a senior scientist at the Alcohol Research Group, wrote that AA indirectly "provides a method for attacking the habits that surround alcohol use. But that is via the people in AA, not the program of AA. The program of AA attacks the base problem, the alcoholic ego, the self-centered, spiritually bereft alcoholic." It is accurate, Kaskutas wrote, that AA provides solutions for alcoholic habits, such as the slogans "go to a meeting if you want to drink," and "avoid slippery people, places, and things." But, Kaskutas wrote, "The slogans aren't the program. The program is the steps. AA aims to go much deeper than addressing the habit part of drinking, and AA founders would argue that attacking the habit is a half measure that won't hold you in good stead; you will eventually succumb to drink unless you change more basic things." For more on the explorations of AA's science, and debates over the program's effectiveness, see C. D. Emrick et al., "Alcoholics Anonymous: What Is Currently Known?" in B. S. McCrady and W. R. Miller, eds., *Research on Alcoholics Anonymous: Opportunities and Alternatives* (New Brunswick, N.J.: Rutgers, 1993), 41–76; John F. Kelly and Mark G. Myers, "Adolescents' Participation in Alcoholics Anonymous and Narcotics Anonymous: Review, Implications, and Future Directions," *Journal of Psychoactive Drugs* 39, no. 3 (September 2007): 259–69; D. R. Groh, L. A. Jason, and C. B. Keys, "Social Network Variables in Alcoholics Anonymous: A Literature Review," *Clinical Psychology Review* 28, no. 3 (March 2008): 430–50; John Francis Kelly, Molly Magill, and Robert Lauren Stout, "How Do People Recover from Alcohol Dependence? A Systematic Review of the Research on Mechanisms of Behavior Change in Alcoholics Anonymous," *Addiction Research and Theory* 17, no. 3 (2009): 236–59.

69 sitting in bed Kurtz, *Not-God.*

69 He chose the number twelve I am indebted to Brendan I. Koerner for his advice, and to his article, "Secret of AA: After 75 Years, We Don't Know How It Works," *Wired*, July 2010; D. R. Davis and G. G. Hansen, "Making Meaning of Alcoholics Anonymous for Social Workers: Myths, Metaphors, and Realities," *Social Work* 43, no. 2 (1998): 169–82.

70 step three, which says Alcoholics Anonymous World Services, *Twelve Steps and Twelve Traditions* (New York: Alcoholics Anonymous World Services, Inc., 2002), 34. Alcoholics Anonymous World Services, *Alcoholics Anonymous: The Big Book*, 4th ed. (New York: Alcoholics Anonymous World Services, Inc., 2002), 59.

70 Because of the program's lack Arthur Cain, "Alcoholics Anonymous: Cult or Cure?" *Harper's Magazine*, February 1963, 48–52; M. Ferri, L. Amato, and M. Davoli, "Alcoholics Anonymous and Other 12-Step Programmes for Alcohol Dependence," *Addiction* 88, no. 4 (1993): 555–62; Harrison M. Trice and Paul Michael Roman, "Delabeling, Relabeling, and Alcoholics Anonymous," *Social Problems* 17, no. 4 (1970): 538–46; Robert E. Tournie, "Alcoholics Anonymous as Treatment and as Ideology," *Journal of Studies on Alcohol* 40, no. 3 (1979): 230–39; P. E. Bebbington, "The Efficacy of Alcoholics Anonymous: The Elusiveness of Hard Data," *British Journal of Psychiatry* 128 (1976): 572–80.

71 **"It's not obvious from the way they're written"** Emrick et al., "Alcoholics Anonymous: What Is Currently Known?"; J. S. Tonigan, R. Toscova, and W. R. Miller, "Meta-analysis of the Literature on Alcoholics Anonymous: Sample and Study Characteristics Moderate Findings," *Journal of Studies on Alcohol* 57 (1995): 65–72; J. S. Tonigan, W. R. Miller, and G. J. Connors, "Project MATCH Client Impressions About Alcoholics Anonymous: Measurement Issues and Relationship to Treatment Outcome," *Alcoholism Treatment Quarterly* 18 (2000): 25–41; J. S. Tonigan, "Spirituality and Alcoholics Anonymous," *Southern Medical Journal* 100, no. 4 (2007): 437–40.

72 **One particularly dramatic demonstration** Heinze et al., "Counteracting Incentive Sensitization in Severe Alcohol Dependence Using Deep Brain Stimulation of the Nucleus Accumbens: Clinical and Basic Science Aspects," *Frontiers in Human Neuroscience* 3, no. 22 (2009).

74 **graduate student named Mandy** "Mandy" is a pseudonym used by the author of the case study this passage draws from.

74 **Mississippi State University** B. A. Dufrene, Steuart Watson, and J. S. Kazmerski, "Functional Analysis and Treatment of Nail Biting," *Behavior Modification* 32 (2008): 913–27.

74 **The counseling center referred Mandy** In a fact-checking letter, the author of this study, Brad Dufrene, wrote that the patient "consented to services at a university-based clinic which was a training and research clinic. At the outset of participating in therapy, she consented to allowing us to use data from her case as in research presentations or publications."

76 **one of the developers of habit reversal training** N. H. Azrin and R. G. Nunn, "Habit-Reversal: A Method of Eliminating Nervous Habits and Tics," *Behaviour Research and Therapy* 11, no. 4 (1973): 619–28; Nathan H. Azrin and Alan L. Peterson, "Habit Reversal for the Treatment of Tourette Syndrome," *Behaviour Research and Therapy* 26, no. 4 (1988): 347–51; N. H. Azrin, R. G. Nunn, and S. E. Frantz, "Treatment of Hairpulling (Trichotillomania): A Comparative Study of Habit Reversal and Negative Practice Training," *Journal of Behavior Therapy and Experimental Psychiatry* 11 (1980): 13–20; R. G. Nunn and N. H. Azrin, "Eliminating Nail-Biting by the Habit Reversal Procedure," *Behaviour Research and Therapy* 14 (1976): 65–67; N. H. Azrin, R. G. Nunn, and S. E. Frantz-Renshaw, "Habit Reversal Versus Negative Practice Treatment of Nervous Tics," *Behavior Therapy* 11, no. 2 (1980): 169–78; N. H. Azrin, R. G. Nunn, and S. E. Frantz-Renshaw, "Habit Reversal Treatment of Thumbsucking," *Behaviour Research and Therapy* 18, no. 5 (1980): 395–99.

77 Today, habit reversal therapy In a fact-checking letter, Dufrene emphasized that methods such as those used with Mandy—known as "simplified habit reversal training"—sometimes differ from other methods of HRT. "My understanding is that Simplified Habit Reversal is effective for reducing habits (e.g., hair pulling, nail biting, thumb sucking), tics (motor and vocal), and stuttering," he wrote. However, other conditions might require more intense forms of HRT. "Effective treatments for depression, smoking, gambling problems, etc. fall under the umbrella term 'Cognitive Behavioral Therapy,'" Dufrene wrote, emphasizing that simplified habit replacement is often not effective for those problems, which require more intensive interventions.

77 verbal and physical tics R. G. Nunn, K. S. Newton, and P. Faucher, "2.5 Years Follow-up of Weight and Body Mass Index Values in the Weight Control for Life! Program: A Descriptive Analysis," *Addictive Behaviors* 17, no. 6 (1992): 579–85; D. J. Horne, A. E. White, and G. A. Varigos, "A Preliminary Study of Psychological Therapy in the Management of Atopic Eczema," *British Journal of Medical Psychology* 62, no. 3 (1989): 241–48; T. Deckersbach et al., "Habit Reversal Versus Supportive Psychotherapy in Tourette's Disorder: A Randomized Controlled Trial and Predictors of Treatment Response," *Behaviour Research and Therapy* 44, no. 8 (2006): 1079–90; Douglas W. Woods and Raymond G. Miltenberger, "Habit Reversal: A Review of Applications and Variations," *Journal of Behavior Therapy and Experimental Psychiatry* 26, no. 2 (1995): 123–31; D. W. Woods, C. T. Wetterneck, and C. A. Flessner, "A Controlled Evaluation of Acceptance and Commitment Therapy Plus Habit Reversal for Trichotillomania," *Behaviour Research and Therapy* 44, no. 5 (2006): 639–56.

78 More than three dozen studies J. O. Prochaska and C. C. DiClemente, "Stages and Processes of Self-Change in Smoking: Toward an Integrative Model of Change," *Journal of Consulting and Clinical Psychology* 51, no. 3 (1983): 390–95; James Prochaska, "Strong and Weak Principles for Progressing from Precontemplation to Action on the Basis of Twelve Problem Behaviors," *Health Psychology* 13 (1994): 47–51; James Prochaska et al., "Stages of Change and Decisional Balance for 12 Problem Behaviors," *Health Psychology* 13 (1994): 39–46; James Prochaska and Michael Goldstein, "Process of Smoking Cessation: Implications for Clinicians," *Clinics in Chest Medicine* 12, no. 4 (1991): 727–35; James O. Prochaska, John Norcross, and Carlo DiClemente, *Changing for Good: A Revolutionary Six-Stage Program for Overcoming Bad Habits and Moving Your Life Positively Forward* (New York: HarperCollins, 1995).

79 "Most of the time, it's not physical" Devin Gordon, "Coach Till You Drop," *Newsweek*, September 2, 2002, 48.

81 during crucial, high-stress moments In fact-checking correspondence, Dungy said he "would not characterize it as falling apart in big games. I would call it not playing well enough in crucial situations, not being able to put those lessons into practice when it was all on the line. St. Louis had one of the highest scoring offenses in the history of the NFL. They managed one TD that game with about 3 minutes left. A team that was scoring almost 38 points a game got 1 TD and 1 FG against the defense, so I hardly think they 'fell apart.'"

81 **"What they were *really* saying"** In fact-checking correspondence, Dungy said "we did lose again in the playoffs to Phil, in another poor showing. This was probably our worst playoff game and it was done under the cloud of rumors, so everyone knew that . . . ownership would be making a coaching change. I think we had instances in the past where we didn't truly trust the system, but I'm not sure that was the case here. Philadelphia was just a tough match-up for us and we couldn't get past them. And not playing well, the score turned out to be ugly. However, it was one of our worst games since the '96 season."

84 **began asking alcoholics** John W. Traphagan, "Multidimensional Measurement of Religiousness/Spirituality for Use in Health Research in Cross-Cultural Perspective," *Research on Aging* 27 (2005): 387–419. Many of those studies use the scale published in G. J. Conners et al., "Measure of Religious Background and Behavior for Use in Behavior Change Research," *Psychology of Addictive Behaviors* 10, no. 2 (June 1996): 90–96.

84 **Then they looked at the data** Sarah Zemore, "A Role for Spiritual Change in the Benefits of 12-Step Involvement," *Alcoholism: Clinical and Experimental Research* 31 (2007): 76s–79s; Lee Ann Kaskutas et al., "The Role of Religion, Spirituality, and Alcoholics Anonymous in Sustained Sobriety," *Alcoholism Treatment Quarterly* 21 (2003): 1–16; Lee Ann Kaskutas et al., "Alcoholics Anonymous Careers: Patterns of AA Involvement Five Years After Treatment Entry," *Alcoholism: Clinical and Experimental Research* 29, no. 11 (2005): 1983–1990; Lee Ann Kaskutas, "Alcoholics Anonymous Effectiveness: Faith Meets Science," *Journal of Addictive Diseases* 28, no. 2 (2009): 145–57; J. Scott Tonigan, W. R. Miller, and Carol Schermer, "Atheists, Agnostics, and Alcoholics Anonymous," *Journal of Studies on Alcohol* 63, no. 5 (2002): 534–54.

87 **Paramedics had rushed him** Jarrett Bell, "Tragedy Forces Dungy 'to Live in the Present,'" *USA Today*, September 1, 2006; Ohm Youngmisuk, "The Fight to Live On," New York *Daily News*, September 10, 2006; Phil Richards, "Dungy: Son's Death Was a 'Test,'" *The Indianapolis Star*, January 25, 2007; David Goldberg, "Tragedy Lessened by Game," *Tulsa World*, January 30, 2007; "Dungy Makes History After Rough Journey," *Akron Beacon Journal*, February 5, 2007; "From Pain, a Revelation," *The New York Times*, July 2007; "Son of Colts' Coach Tony Dungy Apparently Committed Suicide," Associated Press, December 22, 2005; Larry Stone, "Colts Take Field with Heavy Hearts," *The Se-*

attle Times, December 25, 2005; Clifton Brown, "Dungy's Son Is Found Dead; Suicide Suspected," *The New York Times,* December 23, 2005; Peter King, "A Father's Wish," *Sports Illustrated,* February 2007.

88   In a 1994 Harvard study Todd F. Heatherton and Patricia A. Nichols, "Personal Accounts of Successful Versus Failed Attempts at Life Change," *Personality and Social Psychology Bulletin* 20, no. 6 (1994): 664–75.

90   Dungy's team, once again, was I am indebted to Michael Smith, "'Simple' Scheme Nets Big Gains for Trio of Defenses," ESPN.com, December 26, 2005.

90   It's *our* time Michael Silver, "This Time, It's Manning's Moment," *Sports Illustrated,* February 2007.

CHAPTER FOUR

97   They were there to meet For details on O'Neill's life and Alcoa, I am indebted to Paul O'Neill for his generous time, as well as numerous Alcoa executives. I also drew on Pamela Varley, "Vision and Strategy: Paul H. O'Neill at OMB and Alcoa," Kennedy School of Government, 1992; Peter Zimmerman, "Vision and Strategy: Paul H. O'Neill at OMB and Alcoa Sequel," Kennedy School of Government, 1994; Kim B. Clark and Joshua Margolis, "Workplace Safety at Alcoa (A)," *Harvard Business Review,* October 31, 1999; Steven J. Spear, "Workplace Safety at Alcoa (B)," *Harvard Business Review,* December 22, 1999; Steven Spear, *Chasing the Rabbit: How Market Leaders Outdistance the Competition and How Great Companies Can Catch Up and Win* (New York: McGraw-Hill, 2009); Peter Kolesar, "Vision, Values, and Milestones: Paul O'Neill Starts Total Quality at Alcoa," *California Management Review* 35, no. 3 (1993): 133–65; Ron Suskind, *The Price of Loyalty: George W. Bush, the White House, and the Education of Paul O'Neill* (New York: Simon and Schuster, 2004); Michael Arndt, "How O'Neill Got Alcoa Shining," *BusinessWeek,* February 2001; Glenn Kessler, "O'Neill Offers Cure for Workplace Injuries," *The Washington Post,* March 31, 2001; "Pittsburgh Health Initiative May Serve as US Model," Reuters, May 31; S. Smith, "America's Safest Companies: Alcoa: Finding True North," *Occupational Hazards* 64, no. 10 (2002): 53; Thomas A. Stewart, "A New Way to Wake Up a Giant," *Fortune,* October 1990; "O'Neill's Tenure at Alcoa Mixed," Associated Press, December 21, 2000; Leslie Wayne, "Designee Takes a Deft Touch and a Firm Will to Treasury," *The New York Times,* January 16, 2001; Terence Roth, "Alcoa Had Loss of $14.7 Million in 4th Quarter," *The Wall Street Journal,* January 21, 1985; Daniel F. Cuff, "Alcoa Hedges Its Bets, Slowly," *The New York Times,* October 24, 1985; "Alcoa Is Stuck as Two Unions Reject Final Bid," *The Wall Street Journal,* June 2, 1986; Mark Russell, "Alcoa Strike Ends as Two Unions Agree to Cuts in Benefits and to Wage Freezes," *The Wall Street Journal,* July 7, 1986; Thomas F. O'Boyle and Peter Pae, "The Long View: O'Neill Recasts Alcoa with His Eyes Fixed on the Decade Ahead," *The Wall Street Journal*, April 9, 1990; Tracey E. Benson, "Paul O'Neill: True Innovation, True Values, True Leadership," *Industry Week* 242, no. 8 (1993): 24; Joseph Kahn, "Industrialist with a Twist," *The New York Times*, December 21, 2000.

102 O'Neill was one Michael Lewis, "O'Neill's List," *The New York Times*, January 123, 2002; Ron Suskind, *The Price of Loyalty: George W. Bush, the White House, and the Education of Paul O'Neill* (New York: Simon and Schuster, 2004).

103 What mattered was erecting In a fact-checking conversation, O'Neill made clear that the comparison between organizational routines and individual habits is one that he understands and agrees with, but did not explicitly occur to him at the time. "I can relate to that, but I don't own that idea," he told me. Then, as now, he recognizes routines such as the hospital-building program, which is known as the Hill-Burton Act, as an outgrowth of a pattern. "The reason they kept building was because the political instincts are still there that bringing money back home to the district is how people think they get reelected, no matter how much overcapacity we were creating," he told me.

103 "Routines are the organizational analogue" Geoffrey M. Hodgson, "The Nature and Replication of Routines," unpublished manuscript, University of Hertfordshire, 2004, http://www.gredeg.cnrs.fr/routines/workshop/papers/Hodgson.pdf.

104 It became an organizational In a fact-checking conversation, O'Neill wanted to stress that these examples of NASA and the EPA, though illustrative, do not draw on his insights or experiences. They are independently reported.

104 When lawyers asked for permission Karl E. Weick, "Small Wins: Redefining the Scale of Social Problems," *American Psychologist* 39 (1984): 40–49.

104 By 1975, the EPA was issuing http://www.epa.gov/history/topics/epa/15b.htm.

106 He instituted an automatic routine In a fact-checking conversation, O'Neill stressed that he believes that promotions and bonuses should not be tied to worker safety, any more than they should be tied to honesty. Rather, safety is a value that every Alcoa worker should embrace, regardless of the rewards. "It's like saying, 'We're going to pay people more if they don't lie,' which suggests that it's okay to lie a little bit, because we'll pay you a little bit less," he told me. However, it is important to note that in interviews with other Alcoa executives from this period, they said it was widely known that promotions were available only to those employees who evidenced a commitment to safety, and that promise of promotion served as a reward, even if that was not O'Neill's intention.

106 Any time someone was injured In a fact-checking conversation, O'Neill made clear that, at the time, the concept of the "habit loop" was unknown to him. He didn't necessarily think of these programs as fulfilling a criterion for habits, though in retrospect, he acknowledges how his efforts are aligned with more recent research indicating how organizational habits emerge.

108 Take, for instance, studies from P. Callaghan, "Exercise: A Neglected Intervention in Mental Health Care?" *Journal of Psychiatric and Mental Health Nursing* 11 (2004): 476–83; S. N. Blair, "Relationships Between Exercise or Physical Activity and Other Health Behaviors," *Public Health Reports* 100 (2009): 172–80; K. J. Van Rensburg, A. Taylor, and T. Hodgson, "The Effects of Acute Exercise on Attentional Bias Toward Smoking-Related Stimuli During Temporary Abstinence from Smoking," *Addiction* 104, no. 11 (2009): 1910–17; E. R. Ropelle et al., "IL-6 and IL-10 Anti-inflammatory Activity Links Exercise to Hypothalamic Insulin and Leptin Sensitivity Through IKKb and ER Stress Inhibition," *PLoS Biology* 8, no. 8 (2010); P. M. Dubbert, "Physical Activity and Exercise: Recent Advances and Current Challenges," *Journal of Consulting and Clinical Psychology* 70 (2002): 526–36; C. Quinn, "Training as Treatment," *Nursing Standard* 24 (2002): 18–19.

109 Studies have documented that families S. K. Hamilton and J. H. Wilson, "Family Mealtimes: Worth the Effort?" *Infant, Child, and Adolescent Nutrition* 1 (2009): 346–50; American Dietetic Association, "Eating Together as a Family Creates Better Eating Habits Later in Life," ScienceDaily.com, September 4, 2007, accessed April 1, 2011.

109 Making your bed every morning Richard Layard, *Happiness: Lessons from a New Science* (New York: Penguin Press, 2005); Daniel Nettle, *Happiness: The Science Behind Your Smile* (Oxford: Oxford University Press, 2005); Marc Ian Barasch, *Field Notes on the Compassionate Life: A Search for the Soul of Kindness* (Emmaus, Penn.: Rodale, 2005); Alfie Kohn, *Unconditional Parenting: Moving from Rewards and Punishments to Love and Reason* (New York: Atria Books, 2005); P. Alex Linley and Stephen Joseph, eds., *Positive Psychology in Practice* (Hoboken, N.J.: Wiley, 2004).

110 By 7 A.M., I am indebted to the time and help of Bob Bowman in understanding Phelps's training, as well as to Michael Phelps and Alan Abrahamson, *No Limits: The Will to Succeed* (New York: Free Press, 2009); Michael Phelps and Brian Cazeneuve, *Beneath the Surface* (Champaign, Ill.: Sports Publishing LLC, 2008); Bob Schaller, *Michael Phelps: The Untold Story of a Champion* (New York: St. Martin's Griffin, 2008); Karen Crouse, "Avoiding the Deep End When It Comes to Jitters," *The New York Times*, July 26, 2009; Mark Levine, "Out There," *The New York Times*, August 3, 2008; Eric Adelson, "And After That, Mr. Phelps Will Leap a Tall Building in a Single Bound," ESPN .com, July 28, 2008; Sean Gregory, "Michael Phelps: A Real GOAT," *Time*, August 13, 2008; Norman Frauenheim, "Phelps Takes 4th, 5th Gold Medals," *The Arizona Republic*, August 12, 2008.

112 "Once a small win has been accomplished" Karl E. Weick, "Small Wins: Redefining the Scale of Social Problems," *American Psychologist* 39 (1984): 40–49.

112 Small wins fuel transformative changes "Small Wins—The Steady Application of a Small Advantage," Center for Applied Research, 1998, accessed June 24, 2011, http://www.cfar.com/Documents/Smal_win.pdf.

112 It seemed like the gay community's For more details on this incident, see Alix Spiegel's wonderful "81 Words," broadcast on *This American Life*, January 18, 2002, http://www.thisamericanlife.org/.

113 HQ 71-471 ("Abnormal Sexual Relations, Including Sexual Crimes") Malcolm Spector and John I. Kitsuse, *Constructing Social Problems* (New Brunswick, N.J.: Transaction Publishers, 2001).

114 He couldn't tell if they were leaking Phelps and Abrahamson, *No Limits*.

115 It was one additional victory For further discussion of habits and Olympic swimmers, see Daniel Chambliss, "The Mundanity of Excellence," *Sociological Theory* 7 (1989): 70–86.

116 He was killed instantly Paul O'Neill keynote speech, June 25, 2002, at the Juran Center, Carlson School of Management, University of Minnesota, Minneapolis.

118 Rural areas, in particular "Infant Mortality Rates, 1950–2005," http://www.infoplease.com/ipa/A0779935.html; William H. Berentsen, "German Infant Mortality 1960–1980," *Geographical Review* 77 (1987): 157–70; Paul Norman et al., "Geographical Trends in Infant Mortality: England and Wales, 1970–2006," *Health Statistics Quarterly* 40 (2008): 18–29.

119 Today, the U.S. infant mortality World Bank, World Development Indicators. In an email sent in response to fact-checking questions, O'Neill wrote: "This is correct, but I would not take credit for our society doing a better job in reducing infant mortality."

120 They began diets and joined gyms T. A. Wadden, M. L. Butryn, and C. Wilson, "Lifestyle Modification for the Management of Obesity," *Gastroenterology* 132 (2007): 2226–38.

120 Then, in 2009 a group of researchers J. F. Hollis et al., "Weight Loss During the Intensive Intervention Phase of the Weight-Loss Maintenance Trial," *American Journal of Preventative Medicine* 35 (2008): 118–26. See also L. P. Svetkey et al., "Comparison of Strategies for Sustaining Weight Loss, the Weight Loss Maintenance Randomized Controlled Trial," *JAMA* 299 (2008): 1139–48; A. Fitch and J. Bock, "Effective Dietary Therapies for Pediatric

Obesity Treatment," *Reviews in Endocrine and Metabolic Disorders* 10 (2009): 231–36; D. Engstrom, "Eating Mindfully and Cultivating Satisfaction: Modifying Eating Patterns in a Bariatric Surgery Patient," *Bariatric Nursing and Surgical Patient Care* 2 (2007): 245–50; J. R. Peters et al., "Eating Pattern Assessment Tool: A Simple Instrument for Assessing Dietary Fat and Cholesterol Intake," *Journal of the American Dietetic Association* 94 (1994): 1008–13; S. M. Rebro et al., "The Effect of Keeping Food Records on Eating Patterns," *Journal of the American Dietetic Association* 98 (1998): 1163–65.

121 **"After a while, the journal"** For more on weight loss studies, see R. R. Wing and James O. Hill, "Successful Weight Loss Maintenance," *Annual Review of Nutrition* 21 (2001): 323–41; M. L. Klem et al., "A Descriptive Study of Individuals Successful at Long-Term Maintenance of Substantial Weight Loss," *American Journal of Clinical Nutrition* 66 (1997): 239–46; M. J. Mahoney, N. G. Moura, and T. C. Wade, "Relative Efficacy of Self-Reward, Self-Punishment, and Self-Monitoring Techniques for Weight Loss," *Journal of Consulting and Clinical Psychology* 40 (1973): 404–7; M. J. Franz et al., "Weight Loss Outcomes: A Systematic Review and Meta-Analysis of Weight-Loss Clinical Trials with a Minimum 1-Year Follow-up," *Journal of the American Dietetic Association* 107 (2007): 1755–67; A. DelParigi et al., "Successful Dieters Have Increased Neural Activity in Cortical Areas Involved in the Control of Behavior," *International Journal of Obesity* 31 (2007): 440–48.

124 **researchers referred to as "grit"** Jonah Lehrer, "The Truth About Grit," *The Boston Globe*, August 2, 2009.

124 **"despite failure, adversity, and plateaus in progress"** A. L. Duckworth et al., "Grit: Perseverance and Passion for Long-Term Goals," *Journal of Personality and Social Psychology* 92 (2007): 1087–1101.

CHAPTER FIVE

131 **willpower is the single most important** J. P. Tangney, R. F. Baumeister, and A. L. Boone, "High Self-Control Predicts Good Adjustment, Less Pathology, Better Grades, and Interpersonal Success," *Journal of Personality* 72, no. 2 (2004): 271–324; Paul Karoly, "Mechanisms of Self-Regulation: A Systems View," *Annual Review of Psychology* 44 (1993): 23–52; James J. Gross, Jane M. Richards, and Oliver P. John, "Emotional Regulation in Everyday Life," in *Emotion Regulation in Families: Pathways to Dysfunction and Health*, ed. Douglas K. Snyder, Jeffry A. Simpson, and Jan N. Hughes (Washington, D.C.: American Psychological Association, 2006); Katleen De Stobbeleir, Susan Ashford, and Dirk Buyens, "From Trait and Context to Creativity at Work: Feedback-Seeking Behavior as a Self-Regulation Strategy for Creative Performance," Vlerick Leuven Gent Working Paper Series, September 17, 2008; Babette Raabe, Michael Frese, and Terry A. Beehr, "Action Regulation Theory and Career Self-Management," *Journal of Vocational Behavior* 70 (2007): 297–311; Albert Bandura, "The Primacy of Self-Regulation in Health Promotion," *Applied Psychology* 54 (2005): 245–54; Robert G. Lord et al., "Self-Regulation at Work," *Annual Review of Psychology* 61 (2010): 543–68; Colette A. Frayne and Gary P. Latham, "Application of Social Learning Theory to Employee Self-Management of Attendance," *Journal of Applied Psychology* 72 (1987): 387–92; Colette Frayne and J. M. Geringer, "Self-Management Training for Improving Job Performance: A Field Experiment Involving Salespeople," *Journal of Applied Psychology* 85 (2000): 361–72.

131   **"Self-discipline has a bigger effect on"** Angela L. Duckworth and Martin E. P. Seligman, "Self-Discipline Outdoes IQ in Predicting Academic Performance of Adolescents," *Psychological Science* 16 (2005): 939–44.

132   **Executives wrote workbooks that** Information on Starbucks training methods is drawn from numerous interviews, as well as the company's training materials. Information on training materials comes from copies provided by Starbucks employees and court records, including the following internal Starbucks documents and training manuals: *Starbucks Coffee Company Partner Guide, U.S. Store Version; Learning Coach Guide; In-Store Learning Coaches Guide; Shift Supervisor Learning Journey; Retail Management Training; Supervisory Skills Facilitator Guide; Supervisory Skills Partner Workbook; Shift Supervisor Training: Store Manager's Planning and Coaches Guide; Managers' Guide: Learning to Lead, Level One and Two; Supervisory Skills: Learning to Lead Facilitators Guide; First Impressions Guide; Store Manager Training Plan/Guide; District Manager Training Plan/Guide; Partner Resources Manual; Values Walk.* In a statement sent in response to fact-checking inquiries, a Starbucks representative wrote: "In reviewing, we felt that your overall theme focuses on emotional intelligence (EQ) and that we attract partners who need development in this area—this is not true holistically. It's important to note that 70 percent of U.S. partners are students and learning in a lot of ways in their life. What Starbucks provides—and partners are inclined to join because of it—is an environment that matches their values, a place to be a part of something bigger (like community), an approach that focuses on problem solving by showing not telling and a successful way to deliver inspired service." The company added that "we'd like to note that as part of our Customer Service Vision, our partners are trusted completely and are empowered to use their best judgment. We believe that this level of trust and empowerment is unique, and that partners rise to the occasion when we treat them with respect."

133   **It was as if the marshmallow-ignoring kids** Harriet Mischel and Walter Mischel, "The Development of Children's Knowledge of Self-Control Strategies," *Child Development* 54 (1983), 603–19; W. Mischel, Y. Shoda, and M. I. Rodriguez, "Delay of Gratification in Children," *Science* 244 (1989): 933–38; Walter Mischel et al., "The Nature of Adolescent Competencies Predicted by Preschool Delay of Gratification," *Journal of Personality and Social Psychology* 54 (1988): 687–96; J. Metcalfe and W. Mischel, "A Hot/Cool-System Analysis of Delay of Gratification: Dynamics of Will Power," *Psychological Review* 106 (1999): 3–19; Jonah Lehrer, "The Secret of Self Control," *The New Yorker*, May 18, 2009.

137 Some have suggested it helps clarify In a fact-checking email, Muraven wrote: "There is research to suggest that marital problems spring from low self-control and that depletion contributes to poor outcomes when couples are discussing tense relationship issues. Likewise, we have found that on days that require more self-control than average, people are more likely to lose control over their drinking. There is also some research that suggests depleted individuals make poorer decisions than nondepleted individuals. These findings may be extended to explain extramarital affairs or mistakes by physicians, but that has not been" directly shown to be a cause-and-effect relationship.

137 "If you use it up too early" Roy F. Baumeister et al., "Ego-Depletion: Is the Active Self a Limited Resource?" *Journal of Personality and Social Psychology* 18 (1998): 130–50; R. F. Baumeister, M. Muraven, and D. M. Tice, "Self-Control as a Limited Resource: Regulatory Depletion Patterns," *Psychological Bulletin* 126 (1998): 247–59; R. F. Baumeister, M. Muraven, and D. M. Tice, "Longitudinal Improvement of Self-Regulation Through Practice: Building Self-Control Strength Through Repeated Exercise," *Journal of Social Psychology* 139 (1999): 446–57; R. F. Baumeister, M. Muraven, and D. M. Tice, "Ego Depletion: A Resource Model of Volition, Self-Regulation, and Controlled Processing," *Social Cognition* 74 (2000): 1252–65; Roy F. Baumeister and Mark Muraven, "Self-Regulation and Depletion of Limited Resources: Does Self-Control Resemble a Muscle?" *Psychological Bulletin* 126 (2000): 247–59; See also M. S. Hagger et al., "Ego Depletion and the Strength Model of Self-Control: A Meta-Analysis," *Psychological Bulletin* 136 (2010): 495–25; R. G. Baumeister, K. D. Vohs, and D. M. Tice, "The Strength Model of Self-Control," *Current Directions in Psychological Science* 16 (2007): 351–55; M. I. Posne and M. K. Rothbart, "Developing Mechanisms of Self-Regulation," *Development and Psychopathology* 12 (2000): 427–41; Roy F. Baumeister and Todd F. Heatherton, "Self-Regulation Failure: An Overview," *Psychological Inquiry* 7 (1996): 1–15; Kathleen D. Vohs et al., "Making Choices Impairs Subsequent Self-Control: A Limited-Resource Account of Decision Making, Self-Regulation, and Active Initiative," *Journal of Personality and Social Psychology* 94 (2008): 883–98; Daniel Romer et al., "Can Adolescents Learn Self-Control? Delay of Gratification in the Development of Control over Risk Taking," *Prevention Science* 11 (2010):

319–30. In a fact-checking email, Muraven wrote: "Our research suggests that people often don't even realize that they are depleted and that the first act of self-control affected them. Instead, exerting self-control causes people to be less willing to work hard on subsequent self-control efforts (ultimately, this is a theory of motivation, not cognition). . . . [E]ven after the most depleting day, people still don't urinate on the floor. Again, this suggests the motivational aspect of the theory—they lack the motivation to force themselves to do things that are less important to them. I realize this may seem like splitting hairs, but it is critical to understand that self-control doesn't fail because the person cannot muster the needed resources. Instead it fails because the effort seems too great for the payoff. Basically, I don't want the next murderer to say that he was depleted so he couldn't control himself."

138 They enrolled two dozen people Megan Oaten and K. Cheng, "Longitudinal Gains in Self-Regulation from Regular Physical Exercise," *Journal of Health Psychology* 11 (2006): 717–33. See also Roy F. Baumeister et al., "Self-Regulation and Personality: How Interventions Increase Regulatory Success, and How Depletion Moderates the Effects of Traits on Behavior," *Journal of Personality* 74 (2006): 1773–1801.

138 So they designed another experiment Megan Oaten and K. Cheng, "Improvements in Self-Control from Financial Monitoring," *Journal of Economic Psychology* 28 (2007): 487–501.

139 fifteen fewer cigarettes each day Roy F. Baumeister et al., "Self-Regulation and Personality."

139 They enrolled forty-five Ibid.

139 Heatherton, a researcher at Dartmouth For a selection of Heatherton's fascinating work, see *Todd F. Heatherton, Ph.D.*, http://www.dartmouth.edu/~heath/#Pubs, last modified June 30, 2009.

139 Many of these schools have dramatically Lehrer, "The Secret of Self Control."

140 A five-year-old who can follow In a fact-checking email, Dr. Heatherton expanded upon this idea: "Exactly how the brain does this is somewhat unclear, although I propose that people develop better frontal control over subcortical reward centers. . . . The repeated practice helps strengthen the 'muscle' (although clearly it is not a muscle; more likely it is better prefrontal cortical control or the development of a strong network of brain regions involved in controlling behavior)." For more information, see Todd F. Heatherton and Dylan D. Wagner, "Cognitive Neuroscience of Self-Regulation Failure," *Trends in Cognitive Sciences* 15 (2011): 132–39.

140 They sponsored weight-loss classes In a fact-checking email, a Starbucks spokesman wrote: "Currently, Starbucks offers discounts at many of the national fitness clubs. We believe that this discussion should be more around overall health and wellness options provided to our partners, rather than focusing specifically on gym memberships. We know that our partners want to find ways to be well and we continue to look for programs that will enable them to do that."

141 opening seven new stores every day Michael Herriman et al., "A Crack in the Mug: Can Starbucks Mend It?" *Harvard Business Review,* October 2008.

141 In 1992, a British psychologist Sheina Orbell and Paschal Sheeran, "Motivational and Volitional Processes in Action Initiation: A Field Study of the Role of Implementation Intentions," *Journal of Applied Social Psychology* 30, no. 4 (April 2000): 780–97.

145 An impatient crowd might overwhelm In a fact-checking statement, a Starbucks spokesman wrote: "Overall accurate assessment, however, we would argue that any job is stressful. As mentioned above, one of the key elements of our Customer Service Vision is that every partner owns the customer experience. This empowerment lets partners know that the company trusts them to resolve issues and helps create the confidence to successfully navigate these moments."

145 The company identified specific rewards These details were confirmed with Starbucks employees and executives. In a fact-checking statement, however, a Starbucks spokesman wrote: "This is not accurate." The spokesman declined to provide further details.

145 We *Listen* to the customer In a fact-checking statement, a Starbucks spokesman wrote: "While it is certainly not incorrect or wrong to refer to it, LATTE is no longer part of our formal training. In fact, we are moving away from more prescriptive steps like LATTE and are widening the guardrails to enable store partners to engage in problem solving to address the many unique issues that arise in our stores. This model is very dependent on continual effective coaching by shift supervisors, store, and district managers."

146 Then they practice those plans In a fact-checking statement, a Starbucks spokesman wrote: "Overall accurate assessment—we strive to provide tools and training on both skills and behaviors to deliver world-class customer service to every customer on every visit. We would like to note, however, that similar to LATTE (and for the same reasons), we do not formally use Connect, Discover, Respond."

147 " 'This is better than a visit' " Constance L. Hays, "These Days the Customer Isn't Always Treated Right," *The New York Times,* December 23, 1998.

147 Schultz, the man who built Starbucks Information on Schultz from Adi Ignatius, "We Had to Own the Mistakes," *Harvard Business Review,* July-August 2010; William W. George and Andrew N. McLean, "Howard Schultz: Building

Starbucks Community (A)," *Harvard Business Review*, June 2006; Koehn, Besharov, and Miller, "Starbucks Coffee Company in the 21st Century," *Harvard Business Review*, June 2008; Howard Schultz and Dori Jones Yang, *Pour Your Heart Into It: How Starbucks Built a Company One Cup at a Time* (New York: Hyperion, 1997); Taylor Clark, *Starbucked: A Double Tall Tale of Caffeine, Commerce, and Culture* (New York: Little, Brown, 2007); Howard Behar, *It's Not About the Coffee: Lessons on Putting People First from a Life at Starbucks* (New York: Portfolio Trade, 2009); John Moore, *Tribal Knowledge* (New York: Kaplan, 2006); Bryant Simon, *Everything but the Coffee: Learning About America from Starbucks* (Berkeley: University of California Press, 2009). In a fact-checking statement, a Starbucks spokesman wrote: "Although at a very high level, the overall story is correct, a good portion of the details are incorrect or cannot be verified." That spokesperson declined to detail what was incorrect or provide any clarifications.

149 **Mark Muraven, who was by then** M. Muraven, M. Gagné, and H. Rosman, "Helpful Self-Control: Autonomy Support, Vitality, and Depletion," *Journal of Experimental and Social Psychology* 44, no. 3 (2008): 573–85. See also Mark Muraven, "Practicing Self-Control Lowers the Risk of Smoking Lapse," *Psychology of Addictive Behaviors* 24, no. 3 (2010): 446–52; Brandon J. Schmeichel and Kathleen Vohs, "Self-Affirmation and Self-Control: Affirming Core Values Counteracts Ego Depletion," *Journal of Personality and Social Psychology* 96, no. 4 (2009): 770–82; Mark Muraven, "Autonomous Self-Control Is Less Depleting," *Journal of Research in Personality* 42, no. 3 (2008): 763–70; Mark Muraven, Dikla Shmueli, and Edward Burkley, "Conserving Self-Control Strength," *Journal of Personality and Social Psychology* 91, no. 3 (2006): 524–37; Ayelet Fishbach, "The Dynamics of Self-Regulation," in *11th Sydney Symposium of Social Psychology* (New York: Psychology Press, 2001); Tyler F. Stillman et al., "Personal Philosophy and Personnel Achievement: Belief in Free Will Predicts Better Job Performance," *Social Psychological and Personality Science* 1 (2010): 43–50; Mark Muraven, "Lack of Autonomy and Self-Control: Performance Contingent Rewards Lead to Greater Depletion," *Motivation and Emotion* 31, no. 4 (2007): 322–30.

151 **One 2010 study** This study, as of the time of writing this book, was unpublished and shared with me on the condition its authors would not be revealed. However, further details on employee empowerment studies can be found in C. O. Longenecker, J. A. Scazzero, and T. T. Standfield, "Quality Improvement Through Team Goal Setting, Feedback, and Problem Solving: A Field Experiment," *International Journal of Quality and Reliability Management* 11, no. 4 (1994): 45–52; Susan G. Cohen and Gerald E. Ledford, "The Effectiveness of Self-Managing Teams: A Quasi-Experiment," *Human Relations* 47, no. 1 (1994): 13–43; Ferris, Rosen, and Barnum, *Handbook of Human Resource Management* (Cambridge, Mass.: Blackwell Publishers, 1995); Linda Honold,

"A Review of the Literature on Employee Empowerment," *Empowerment in Organizations* 5, no. 4 (1997): 202–12; Thomas C. Powell, "Total Quality Management and Competitive Advantage: A Review and Empirical Study," *Strategic Management Journal* 16 (1995): 15–37.

CHAPTER SIX

154 Afterward, he had trouble staying awake Details on this case come from a variety of sources, including interviews with the professionals involved, witnesses in the operating room and emergency room, and news accounts and documents published by the Rhode Island Department of Health. Those include consent orders published by the Rhode Island Department of Health; the Statement of Deficiencies and Plan of Correction published by Rhode Island Hospital on August 8, 2007; Felicia Mello, "Wrong-Site Surgery Case Leads to Probe," *The Boston Globe*, August 4, 2007; Felice Freyer, "Doctor to Blame in Wrong-Side Surgery, Panel Says," *The Providence Journal*, October 14, 2007; Felice Freyer, "R.I. Hospital Cited for Wrong-Side Surgery," *The Providence Journal*, August 3, 2007; "Doctor Disciplined for Wrong-Site Brain Surgery," Associated Press, August 3, 2007; Felice Freyer, "Surgeon Relied on Memory, Not CT Scan," *The Providence Journal*, August 24, 2007; Felicia Mello, "Wrong-Site Surgery Case Leads to Probe 2nd Case of Error at R.I. Hospital This Year," *The Boston Globe*, August 4, 2007; "Patient Dies After Surgeon Operates on Wrong Side of Head," Associated Press, August 24, 2007; "Doctor Back to Work After Wrong-Site Brain Surgery," Associated Press, October 15, 2007; Felice Freyer, "R.I. Hospital Fined After Surgical Error," *The Providence Journal*, November 27, 2007.

155 Unless the blood was drained Accounts of this case were described by multiple individuals, and some versions of events differ with one another. Those differences, where appropriate, are described in the notes.

155 In 2002, the National Coalition on Health Care http://www.rhodeisland hospital.org/rih/about/milestones.htm.

155 "They can't take away our pride." Mark Pratt, "Nurses Rally on Eve of Contract Talks," Associated Press, June 22, 2000; "Union Wants More Community Support During Hospital Contract Dispute," Associated Press, June 25, 2000; "Nurses Say Staff Shortage Hurting Patients," Associated Press, August 31, 2000; "Health Department Surveyors Find Hospitals Stressed," Associated Press, November 18, 2001; "R.I. Hospital Union Delivers Strike Notice," Associated Press, June 20, 2000.

155 Administrators eventually agreed to limit In a statement, a spokeswoman for Rhode Island Hospital said: "The strike was not about relationships between physicians and nurses, it was about wages and work rules. Mandatory overtime is a common practice and has been an issue in unionized hospitals across the country. I don't know whether there were signs with those messages during the 2000 union negotiations, but if so, they would have referred to mandatory overtime, not relationships between physicians and nurses."

155 to make sure mistakes are avoided American Academy of Orthopaedic Surgeons Joint Commission Guidelines, http://www3.aaos.org/member/safety/guidelines.cfm.

157 A half hour later RIDH Statement of Deficiencies and Plan of Correction, August 7, 2007.

157 There was no clear indication of In a statement, Rhode Island Hospital said some of these details are incorrect, and referred to the August 7, 2007, RIDH Statement of Deficiencies and Plan of Correction. That document says, "There is no evidence in the medical record that the Nurse Practitioner, employed by the covering Neurosurgeon, received, or attempted to obtain, the necessary information related to the patient's CT scan . . . to confirm the correct side of the bleed and [sic] prior to having the consent form signed for craniotomy surgery. . . . The medical record indicates that the surgical consent was obtained by a Nurse Practitioner working for the Neurosurgeon who was on call. Although the surgical consent indicates that the procedure to be performed was a 'Right craniotomy and evacuation of subdural hematoma,' the side (right) was not initially entered onto the consent form. Interview on 8/2/07 at 2:05 PM with the Director of Perioperative Surgery indicated that patient . . . was transported from the emergency department with an incomplete (as to side) signed surgical consent. The Circulating Nurse noted that the site of the craniotomy was not included on the signed surgical consent as required by hospital policy. She indicated that the site of the craniotomy surgery was then added by the Neurosurgeon, in the operating room, once he was questioned by the Circulating Nurse regarding the site of the surgery." In a follow-up statement, Rhode Island Hospital wrote that the surgeon "and his assistant finished the spinal surgery, the OR was readied, and when they were in the hall, about to return to the OR, the OR nurse saw the consent form did not include the side of the surgery and told [the surgeon]. The doctor took the consent from the nurse and wrote 'right' on it."

158 "We have to operate immediately." In a letter sent in response to fact-checking inquiries, the physician involved in this case contradicted or challenged some of the events described in this chapter. The physician wrote that the nurse in this case was not concerned that the physician was operating on the wrong side. The nurse's concern focused on paperwork issues. The physician contended that the nurse did not question the physician's expertise or accuracy. The nurse did not ask the physician to pull up the films, according to the physician. The physician said that he asked the nurse to find the family to see if it was possible to "redo the consent form properly," rather than the other way around. When the family could not be found, according to the physician, the physician asked for clarification from the nurse regarding the procedure to improve the paperwork. The nurse, according to the physician, said he wasn't sure, and as a result, the physician decided to "put a correction to the consent form and write a note in the chart detailing that we needed to proceed." The physician said he never swore and was not excited.

Rhode Island Hospital, when asked about this account of events, said it was not accurate and referred to the August 7, 2007, RIDH Statement of Deficiencies and Plan of Correction. In a statement, the hospital wrote, "During our investigation, no one said they heard [the surgeon] say that the patient was going to die."

"Those quotes with all the excitement and irritation in my manner, even swearing was completely inaccurate," the physician wrote. "I was calm and professional. I showed some emotion only for a brief moment when I realized I had started on the wrong side. The critical problem was that we would not have films to look at during the procedure. . . . Not having films to view during the case is malpractice by the hospital; however we had no choice but to proceed without films."

Rhode Island Hospital responded that the institution "can't comment on [the surgeon's] statement but would note that the hospital assumed that surgeons would put films up as they performed surgery if there was any question about the case. After this event, the hospital mandated that films would be available for the team to view." In a second statement, the hospital wrote the surgeon "did not swear during this exchange. The nurse told [the surgeon] he had not received report from the ED and the nurse spent several minutes in the room trying to reach the correct person in the ED. The NP indicated he had received report from the ED physician. However, the CRNA (nurse anesthetist) needed to know the drugs that had been given in the ED, so the nurse was going thru the record to get her the info."

The Rhode Island Board of Medical Licensure and Discipline, in a consent order, wrote that the physician "failed to make an accurate assessment of the location of the hematoma prior to performing the surgical evacuation." The State Department of Health found that "an initial review of this incident reveals hospital surgical safeguards are deficient and that some systems were not followed."

Representatives of both the Board and Department of Health declined to comment further.

159   the surgeon yelled In a statement, a representative of Rhode Island Hospital wrote "I believe [the surgeon] was the one who noticed that there was no bleeding—there are various versions as to what he said at that time. He asked for the films to be pulled up, confirmed the error and they proceeded to close and perform the procedure on the correct side. Except for [the surgeon's] comments, the staff said the room was very quiet once they realized the error."

159   ever working at Rhode Island Hospital again In the physician's letter responding to fact-checking inquiries, he wrote that "no one has claimed that this mistake cost [the patient] his life. The family never claimed wrongful death, and they personally expressed their gratitude to me for saving his life on that day. The hospital and the nurse practitioner combined paid more towards a $140,000 settlement than I did." Rhode Island Hospital, when asked about this account, declined to comment.

160   The book's bland cover and daunting R. R. Nelson and S. G. Winter, *An Evolutionary Theory of Economic Change* (Cambridge, Mass.: Belknap Press of Harvard University Press, 1982).

160   candidates didn't pretend to understand R. R. Nelson and S. G. Winter, "The Schumpeterian Tradeoff Revisited," *The American Economic Review* 72 (1982): 114–32. Winter, in a note in response to fact-checking questions, wrote: "The 'Schumpeterian tradeoff' (subject of a 1982 AER paper and a kindred chapter, 14, in our book) was only a facet of the project, and not a motivating one. Nelson and I were discussing a collection of issues around technological change, economic growth and firm behavior long before 1982, long before we were together at Yale, and particularly at RAND in 1966–68. Nelson went to Yale in 1968; I went to Michigan that year and joined the Yale faculty in 1976. We were 'on the trail' of the 1982 book from 1967, and started publishing related work in 1973. . . . In short, while the 'Schumpeter' influence is obviously strong in the heritage, the specific 'Schumpeterian tradeoff' aspect is not."

160   Within the world of business strategy For an overview of subsequent research, see M. C. Becker, "Organizational Routines: A Review of the Literature," *Industrial and Corporate Change* 13 (2004): 643–78; Marta S. Feldman, "Organizational Routines as a Source of Continuous Change," *Organization Science* 11 (2000): 611–29.

160   before arriving at their central conclusion Winter, in a note in response to fact-checking questions, wrote: "There was very little empirical work of my own, and even less that got published—most of that being Nelson on aspects of technological change. In the domain of firm behavior, we mostly stood on the shoulders of the giants of the Carnegie School (Simon, Cyert, and March), and relied on a wide range of other sources—technology studies, business his-tories, development economics, some psychologists . . . and Michael Polanyi, however you classify him."

161 thousands of employees' independent decisions Winter, in a note in response to fact-checking questions, clarified that such patterns that emerge from thousands of employees' independent decisions are an aspect of routines, but routines also "get shaped from a lot of directions, one of which is deliberate managerial design. We emphasized, however, that when that happens, the actual routine that emerges, as opposed to the nominal one that was deliberately designed, is influenced, again, by a lot of choices at the individual level, as well as other considerations (see book [*Evolutionary Theory of Economic Change*] p. 108)."

161 These organizational habits—or "routines" For more on the fascinating topic of how organizational routines emerge and work, see Paul S. Adler, Barbara Goldoftas, and David I. Levine, "Flexibility Versus Efficiency? A Case Study of Model Changeovers in the Toyota Production System," *Organization Science* 10 (1999): 43–67; B. E. Ashforth and Y. Fried, "The Mindlessness of Organisational Behaviors," *Human Relations* 41 (1988): 305–29; Donde P. Ashmos, Dennis Duchon, and Reuben R. McDaniel, "Participation in Strategic Decision Making: The Role of Organisational Predisposition and Issue Interpretation," *Decision Sciences* 29 (1998): 25–51; M. C. Becker, "The Influence of Positive and Negative Normative Feedback on the Development and Persistence of Group Routines," doctoral thesis, Purdue University, 2001; M. C. Becker and N. Lazaric, "The Role of Routines in Organizations: An Empirical and Taxonomic Investigation," doctoral thesis, Judge Institute of Management, University of Cambridge, 2004; Bessant, Caffyn, and Gallagher, "The Influence of Knowledge in the Replication of Routines," *Economie Appliquée* LVI, 65–94; "An Evolutionary Model of Continuous Improvement Behaviour," *Technovation* 21 (2001): 67–77; Tilmann Betsch, Klaus Fiedler, and Julia Brinkmann, "Behavioral Routines in Decision Making: The Effects of Novelty in Task Presentation and Time Pressure on Routine Maintenance and Deviation," *European Journal of Psychology* 28 (1998): 861–78; Tilmann Betsch et al., "When Prior Knowledge Overrules New Evidence: Adaptive Use of Decision Strategies and Role Behavioral Routines," *Swiss Journal of Psychology* 58 (1999): 151–60; Tilmann Betsch et al., "The Effects of Routine Strength on Adaptation and Information Search in Recurrent Decision Making," *Organisational Behaviour and Human Decision Processes* 84 (2001): 23–53; J. Burns, "The Dynamics of Accounting Change: Interplay Between New Practices, Routines, Institutions, Power, and Politics," *Accounting, Auditing and Accountability Journal* 13 (2000): 566–86; M. D. Cohen, "Individual Learning and Organisational Routine: Emerging Connections," *Organisation Science* 2 (1991): 135–39; M. Cohen and P. Bacdayan, "Organisational Routines Are Stored as Procedural

Memory: Evidence from a Laboratory Study," *Organisation Science* 5 (1994): 554–68; M. D. Cohen et al., "Routines and Other Recurring Action Patterns of Organisations: Contemporary Research Issues," *Industrial and Corporate Change* 5 (1996): 653–98; B. Coriat, "Variety, Routines, and Networks: The Metamorphosis of Fordist Firms," *Industrial and Corporate Change* 4 (1995): 205–27; B. Coriat and G. Dosi, "Learning How to Govern and Learning How to Solve Problems: On the Co-evolution of Competences, Conflicts, and Organisational Routines," in *The Role of Technology, Strategy, Organisation, and Regions*, ed. A. D. J. Chandler, P. Hadstroem, and O. Soelvell (Oxford: Oxford University Press, 1998); L. D'Adderio, "Configuring Software, Reconfiguring Memories: The Influence of Integrated Systems on the Reproduction of Knowledge and Routines," *Industrial and Corporate Change* 12 (2003): 321–50; P. A. David, *Path Dependence and the Quest for Historical Economics: One More Chorus of the Ballad of QWERTY* (Oxford: Oxford University Press, 1997); G. Delmestri, "Do All Roads Lead to Rome . . . or Berlin? The Evolution of Intra- and Inter-organisational Routines in the Machine-Building Industry," *Organisation Studies* 19 (1998): 639–65; Giovanni Dosi, Richard R. Nelson, and Sidney Winter, "Introduction: The Nature and Dynamics of Organisational Capabilities," *The Nature and Dynamics of Organisational Capabilities*, ed. G. Dosi, R. R. Nelson, and S. G. Winter (Oxford: Oxford University Press, 2000), 1–22; G. Dowell and A. Swaminathan, "Racing and Back-pedalling into the Future: New Product Introduction and Organisational Mortality in the US Bicycle Industry, 1880–1918," *Organisation Studies* 21 (2000): 405–31; A. C. Edmondson, R. M. Bohmer, and G. P. Pisano, "Disrupted Routines: Team Learning and New Technology Implementation in Hospitals," *Administrative Science Quarterly* 46 (2001): 685–716; M. Egidi, "Routines, Hierarchies of Problems, Procedural Behaviour: Some Evidence from Experiments," in *The Rational Foundations of Economic Behaviour*, ed. K. Arrow et al. (London: Macmillan, 1996), 303–33; M. S. Feldman, "Organisational Routines as a Source of Continuous Change," *Organisation Science* 11 (2000): 611–29; Marta S. Feldman, "A Performative Perspective on Stability and Change in Organizational Routines," *Industrial and Corporate Change* 12 (2003): 727–52; Marta S. Feldman and B. T. Pentland, "Reconceptualizing Organizational Routines as a Source of Flexibility and Change," *Administrative Science Quarterly* 48 (2003): 94–118; Marta S. Feldman and A. Rafaeli, "Organisational Routines as Sources of Connections and Understandings," *Journal of Management Studies* 39 (2002): 309–31; A. Garapin and A. Hollard, "Routines and Incentives in Group Tasks," *Journal of Evolutionary Economics* 9 (1999): 465–86; C. J. Gersick and J. R. Hackman, "Habitual Routines in Task-Performing Groups," *Organisational Behaviour and Human Decision Processes* 47 (1990): 65–97; R. Grant, "Toward a Knowledge-Based Theory of the Firm," *Strategic Management Journal* 17

(1996): 109–22; R. Heiner, "The Origin of Predictable Behaviour," *American Economic Review* 73 (1983): 560–95; G. M. Hodgson, "The Ubiquity of Habits and Rules," *Cambridge Journal of Economics* 21 (1997): 663–84; G. M. Hodgson, "The Mystery of the Routine: The Darwinian Destiny of *An Evolutionary Theory of Economic Change*," *Revue Économique* 54 (2003): 355–84; G. M. Hodgson and T. Knudsen, "The Firm as an Interactor: Firms as Vehicles for Habits and Routines," *Journal of Evolutionary Economics* 14, no. 3 (2004): 281–307; A. Inam, "Institutions, Routines, and Crises: Post-earthquake Housing Recovery in Mexico City and Los Angeles," doctoral thesis, University of Southern California, 1997; A. Inam, "Institutions, Routines, and Crises—Post-earthquake Housing Recovery in Mexico City and Los Angeles," *Cities* 16 (1999): 391–407; O. Jones and M. Craven, "Beyond the Routine: Innovation Management and the Teaching Company Scheme," *Technovation* 21 (2001): 267–79; M. Kilduff, "Performance and Interaction Routines in Multinational Corporations," *Journal of International Business Studies* 23 (1992): 133–45; N. Lazaric, "The Role of Routines, Rules, and Habits in Collective Learning: Some Epistemological and Ontological Considerations," *European Journal of Economic and Social Systems* 14 (2000): 157–71; N. Lazaric and B. Denis, "How and Why Routines Change: Some Lessons from the Articulation of Knowledge with ISO 9002 Implementation in the Food Industry," *Economies et Sociétés* 6 (2001): 585–612; B. Levitt and J. March, "Organisational Learning," *Annual Review of Sociology* 14 (1988): 319–40; P. Lillrank, "The Quality of Standard, Routine, and Nonroutine Processes," *Organization Studies* 24 (2003): 215–33; S. Massini et al., "The Evolution of Organizational Routines Among Large Western and Japanese Firms," *Research Policy* 31 (2002): 1333–48; T. J. McKeown, "Plans and Routines, Bureaucratic Bargaining, and the Cuban Missile Crisis," *Journal of Politics* 63 (2001): 1163–90; A. P. Minkler, "The Problem with Dispersed Knowledge: Firms in Theory and Practice," *Kyklos* 46 (1993): 569–87; P. Morosini, S. Shane, and H. Singh, "National Cultural Distance and Cross-Border Acquisition Performance," *Journal of International Business Studies* 29 (1998): 137–58; A. Narduzzo, E. Rocco, and M. Warglien, "Talking About Routines in the Field," in *The Nature and Dynamics of Organizational Capabilities*, ed. G. Dosi, R. Nelson, and S. Winter (Oxford: Oxford University Press, 2000), 27–50; R. R. Nelson, "Routines," in *The Elgar Companion to Institutional and Evolutionary Economics*, vol. 2, ed. G. Hodgson, W. Samuels, and M. Tool (Aldershot, U.K.: Edward Elgar, 1992), 249–53; B. T. Pentland, "Conceptualizing and Measuring Variety in the Execution of Organizational Work Processes," *Management Science* 49 (2003): 857–70; B. T. Pentland and H. Rueter, "Organisational Routines as Grammars of Action," *Administrative Sciences Quarterly* 39 (1994): 484–510; L. Perren and P. Grant, "The Evolution of Management Accounting Routines in Small Businesses: A Social Construction

Perspective," *Management Accounting Research* 11 (2000): 391–411; D. J. Phillips, "A Genealogical Approach to Organizational Life Chances: The Parent–Progeny Transfer Among Silicon Valley Law Firms, 1946–1996," *Administrative Science Quarterly* 47 (2002): 474–506; S. Postrel and R. Rumelt, "Incentives, Routines, and Self-Command," *Industrial and Corporate Change* 1 (1992): 397–425; P. D. Sherer, N. Rogovksy, and N. Wright, "What Drives Employment Relations in Taxicab Organisations?" *Organisation Science* 9 (1998): 34–48; H. A. Simon, "Programs as Factors of Production," *Proceedings of the Nineteenth Annual Winter Meeting, 1966*, Industrial Relations Research Association, 1967, 178–88; L. A. Suchman, "Office Procedure as Practical Action: Models of Work and System Design," *ACM Transactions on Office Information Systems* 1 (1983): 320–28; G. Szulanski, "Appropriability and the Challenge of Scope: Banc One Routinizes Replication," in *Nature and Dynamics of Organisational Capabilities*, ed. G. Dosi, R. R. Nelson, and S. G. Winter (Oxford: Oxford University Press, 1999), 69–97; D. Tranfield and S. Smith, "The Strategic Regeneration of Manufacturing by Changing Routines," *International Journal of Operations and Production Management* 18 (1998): 114–29; Karl E. Weick, "The Vulnerable System: An Analysis of the Tenerife Air Disaster," *Journal of Management* 16 (1990): 571–93; Karl E. Weick, "The Collapse of Sensemaking in Organizations: The Mann–Gulch Disaster," *Administrative Science Quarterly* 38 (1993): 628–52; H. M. Weiss and D. R. Ilgen, "Routinized Behaviour in Organisations," *Journal of Behavioral Economics* 14 (1985): 57–67; S. G. Winter, "Economic 'Natural Selection' and the Theory of the Firm," *Yale Economic Essays* 4 (1964): 225–72; S. G. Winter, "Optimization and Evolution in the Theory of the Firm," in *Adaptive Economic Models*, ed. R. Day and T. Groves (New York: Academic Press, 1975), 73–118; S. G. Winter and G. Szulanski, "Replication as Strategy," *Organization Science* 12 (2001): 730–43; S. G. Winter and G. Szulanski, "Replication of Organisational Routines: Conceptualizing the Exploitation of Knowledge Assets," in *The Strategic Management of Intellectual Capital and Organisational Knowledge: A Collection of Readings*, ed. N. Bontis and C. W. Choo (New York: Oxford University Press, 2001), 207–21; M. Zollo, J. Reuer, and H. Singh, "Interorganizational Routines and Performance in Strategic Alliances," *Organization Science* 13 (2002): 701–13.

161 **hundreds of unwritten rules** Esbjoern Segelod, "The Content and Role of the Investment Manual: A Research Note," *Management Accounting Research* 8, no. 2 (1997): 221–31; Anne Marie Knott and Bill McKelvey, "Nirvana Efficiency: A Comparative Test of Residual Claims and Routines," *Journal of Economic Behavior and Organization* 38 (1999): 365–83; J. H. Gittell, "Coordinating Mechanisms in Care Provider Groups: Relational Coordination as a Mediator and Input Uncertainty as a Moderator of Performance Effects," *Management Science* 48 (2002): 1408–26; A. M. Knott and Hart Posen, "Firm

R&D Behavior and Evolving Technology in Established Industries," *Organization Science* 20 (2009): 352–67.

161 companies need to operate G. M. Hodgson, *Economics and Evolution* (Cambridge: Polity Press, 1993); Richard N. Langlois, "Transaction-Cost Economics in Real Time," *Industrial and Corporate Change* (1992): 99–127; R. R. Nelson, "Routines"; R. Coombs and J. S. Metcalfe, "Organizing for Innovation: Co-ordinating Distributed Innovation Capabilities," in *Competence, Governance, and Entrepreneurship*, ed. J. N. Foss and V. Mahnke (Oxford: Oxford University Press, 2000); R. Amit and M. Belcourt, "HRM Processes: A Value-Creating Source of Competitive Advantage," *European Management Journal* 17 (1999): 174–81.

161 They provide a kind of "organizational memory" G. Dosi, D. Teece, and S. G. Winter, "Toward a Theory of Corporate Coherence: Preliminary Remarks," in *Technology and Enterprise in a Historical Perspective*, ed. G. Dosi, R. Giannetti, and P. A. Toninelli (Oxford: Clarendon Press, 1992), 185–211; S. G. Winter, Y. M. Kaniovski, and G. Dosi, "A Baseline Model of Industry Evolution," *Journal of Evolutionary Economics* 13, no. 4 (2003): 355–83; B. Levitt and J. G. March, "Organizational Learning," *Annual Review of Sociology* 14 (1988): 319–40; D. Teece and G. Pisano, "The Dynamic Capabilities of Firms: An Introduction," *Industrial and Corporate Change* 3 (1994): 537–56; G. M. Hodgson, "The Approach of Institutional Economics," *Journal of Economic Literature* 36 (1998): 166–92; Phillips, "Genealogical Approach to Organizational Life Chances"; M. Zollo, J. Reuer, and H. Singh, "Interorganizational Routines and Performance in Strategic Alliances," *Organization Science* 13 (2002): 701–13; P. Lillrank, "The Quality of Standard, Routine, and Nonroutine Processes," *Organization Studies* 24 (2003): 215–33.

162 Routines reduce uncertainty M. C. Becker, "Organizational Routines: A Review of the Literature," *Industrial and Corporate Change* 13, no. 4 (2004): 643–78.

162 But among the most important benefits B. Coriat and G. Dosi, "Learning How to Govern and Learning How to Solve Problems: On the Co-evolution of Competences, Conflicts, and Organisational Routines," in *The Role of Technology, Strategy, Organisation, and Regions*, ed. A. D. J. Chandler, P. Hadstroem, and O. Soelvell (Oxford: Oxford University Press, 1998); C. I. Barnard, *The Functions of the Executive* (Cambridge, Mass.: Harvard University Press, 1938); P. A. Mangolte, "La dynamique des connaissances tacites et articulées: une approche socio-cognitive," *Economie Appliquée* 50, no. 2 (1997): 105–34; P. A. Mangolte, "Le concept de 'routine organisationelle' entre cognition et institution," doctoral thesis, Université Paris-Nord, U.F.R. de Sciences Economiques et de Gestion, Centre de Recherche en Economie Industrielle, 1997; P. A. Mangolte, "Organisational Learning and the Organisational Link: The Problem of

Conflict, Political Equilibrium and Truce," *European Journal of Economic and Social Systems* 14 (2000): 173–90; N. Lazaric and P. A. Mangolte, "Routines et mémoire organisationelle: un questionnement critique de la perspective cognitiviste," *Revue Internationale de Systémique* 12 (1998): 27–49; N. Lazaric and B. Denis, "How and Why Routines Change: Some Lessons from the Articulation of Knowledge with ISO 9002 Implementation in the Food Industry," *Economies et Sociétés* 6 (2001): 585–612; N. Lazaric, P. A. Mangolte, and M. L. Massué, "Articulation and Codification of Know-How in the Steel Industry: Some Evidence from Blast Furnace Control in France," *Research Policy* 32 (2003): 1829–47; J. Burns, "The Dynamics of Accounting Change: Interplay Between New Practices, Routines, Institutions, Power, and Politics," *Accounting, Auditing and Accountability Journal* 13 (2000): 566–86.

163   you'll probably get taken care of over time Winter, in a note in response to fact-checking questions, wrote: "The 'routine as truce' formulation has turned out to have particularly long legs, and I think that is because anybody with some experience in working inside an organization quickly recognizes it as a convenient label for the sorts of goings-on that they are very familiar with. . . . But some of your example about the salesperson evokes issues of trust, cooperation, and organizational culture that go beyond the scope of 'routine as truce.' Those are subtle issues, which can be illuminated from a variety of directions. The 'routine as truce' idea is a lot more specific than related ideas about 'culture.' It says, 'If you, Mr. or Ms. Manager, VISIBLY DEFECT from a widely shared understanding of 'how we do things around here,' you are going to encounter strong resistance, fueled by levels of suspicion about your motives that are far beyond anything you might reasonably expect. And if these responses are not entirely independent of the quality of the arguments you advance, they will be so nearly independent that you will find it hard to see any difference.' So, for example, suppose we take your 'red this year' example down the road a bit, into the implementation phase, where enormous effort has gone into making sure that the red on the sweater is the same on the catalog cover and on catalog p. 17, and both of those match what is in the CEO's head, and that red is also the same one produced in response to contracts with suppliers in Malaysia, Thailand, and Guatemala. That stuff is at the other end of the routines spectrum from the decision on 'red'; people are engaged in complex coordinated behavior—it is more like the semiconductor case. People in the organization think they know what they are doing (because they did more or less the same with the green pullovers featured last year), and they are working like hell to do it, more or less on time. This is guts management stuff, and it is very hard work, thanks partly, in this case, to the (alleged) fact that the human eye can distinguish 7 million different colors. Into that, YOU, Mr. or Ms. Manager, come in and say 'Sorry, it's a mistake, it should be purple. I know we are well down the road with our commitment to red, but hear me out, because . . .' If you have lined up strong allies in the organization who also favor a belated switch to purple, you have just touched off another battle in the 'civil war,' with uncertain consequence. If you don't have such allies, your espoused cause and you are both dead in the organization, in short order. And it doesn't matter what logic and evidence you offer following your 'because.'"

163   of throwing a rival overboard" Nelson and Winter, *Evolutionary Theory of Economic Change,* 110.

165   But that's not enough Rik Wenting, "Spinoff Dynamics and the Spatial Formation of the Fashion Design Industry, 1858–2005," *Journal of Economic Geography* 8, no. 5 (2008): 593–614. Wenting, in a response to fact-checking questions, wrote: "Nelson and Winter speak of organisational routines as repetitive collective actions which determine firm behaviour and performance. Notably they argue that routines are hard to codify and part of company culture, and as such are hard to change. Also, routines are a major reason why firms differ in their performance and the continued difference over time between firms. The literature started by Steven Klepper interpreted this aspect of routines as part of the reason why spinoffs are in performance similar to their parents. I use this same reasoning in the fashion design industry: fashion design entrepreneurs form to a large extent their new firm's blueprint based on the organisational routines learned at their former employer. In my PhD research, I found evidence that from the start of the haute couture industry (1858 Paris), spinoff designer firms (whether located in NY, Paris, Milan or London, etc.) do indeed have a similar performance as their motherfirms."

165   and found the right alliances Details regarding truces—as opposed to routines—within the fashion industry draw on interviews with designers themselves. Wenting, in a response to fact-checking questions, wrote: "Note that I do not speak of truces between entrepreneur and former employer. This is an extension of the organisational routines literature I did not specifically explore. However, in my research on the 'inheritage' effect between motherfirm and spinoff, the role of 'reputation' and 'social network' are often times mentioned by designers in how they experience advantages of their mother company."

166   Philip Brickell, a forty-three-year-old Rodney Cowton and Tony Dawe, "Inquiry Praises PC Who Helped to Fight King's Cross Blaze," *The Times,* February 5, 1988.

166   at the bottom of a nearby escalator Details on this incident come from a variety of sources, including interviews, as well as D. Fennell, *Investigation into the King's Cross Underground Fire* (Norwich, U.K.: Stationery Office Books,

1988); P. Chambers, *Body 115: The Story of the Last Victim of the King's Cross Fire* (New York: John Wiley and Sons, 2006); K. Moodie, "The King's Cross Fire: Damage Assessment and Overview of the Technical Investigation," *Fire Safety Journal* 18 (1992): 13–33; A. F. Roberts, "The King's Cross Fire: A Correlation of the Eyewitness Accounts and Results of the Scientific Investigation," *Fire Safety Journal*, 1992; "Insight: Kings Cross," *The Sunday Times*, November 22, 1987; "Relatives Angry Over Tube Inquest; King's Cross Fire," *The Times*, October 5, 1988.

169   if they aren't designed just right In the Fennell report, the investigator was ambiguous about how much of the tragedy could have been averted if the burning tissue had been reported. The Fennell report is deliberately agnostic about this point: "It will remain a matter of conjecture what would have happened if the London Fire Brigade had been summoned to deal with the burning tissue. . . . It is a matter of speculation what course things would have taken if he had followed the new procedure and called the London Fire Brigade immediately."

173   "Why didn't someone take charge?" "Answers That Must Surface—The King's Cross Fire Is Over but the Controversy Continues," *The Times*, December 2, 1987; "Businessman Praised for Rescuing Two from Blazing Station Stairwell; King's Cross Fire Inquest," *The Times*, October 6, 1998.

175   responsibility for passengers' safety In a statement in response to fact-checking questions, a spokesman for London Underground and Rail wrote: "London Underground has given this careful consideration and will not, on this occasion, be able to provide further comment or assistance on this. LU's response to the King's Cross fire and the organisational changes made to address the issues are well-documented, and the sequence of events leading to the fire is covered in great detail in Mr Fennell's report, so LU does not consider it necessary to add more comment to the already large body of work on the matter. I appreciate this is not the response you were hoping for."

176   the hospital was fined another $450,000 Felice Freyer, "Another Wrong-Site Surgery at R.I. Hospital," *The Providence Journal*, October 28, 2009; "Investigators Probing 5th Wrong-Site Surgery at Rhode Island Hospital Since 2007," Associated Press, October 23, 2009; "R.I. Hospital Fined $150,000 in 5th Wrong-Site Surgery Since 2007, Video Cameras to Be Installed," Associated Press, November 2, 2009; Letter to Rhode Island Hospital from Rhode Island Department of Health, November 2, 2009; Letter to Rhode Island Hospital from Rhode Island Department of Health, October 26, 2010; Letter to Rhode Island Hospital from Centers for Medicare and Medicaid Services, October 25, 2010.

176   "The problem's not going away," "'The Problem's Not Going Away': Mistakes Lead to Wrong-Side Brain Surgeries at R.I. Hospital," Associated Press, December 15, 2007.

176 **"everything was out of control."** In a statement, a Rhode Island Hospital spokeswoman wrote: "I never heard of any reporter 'ambushing' a doctor—and never saw any such incident on any of the news stations. While I can't comment on individual perceptions, the quote implies a media frenzy, which did not happen. While the incidents received national attention, none of the national media came to Rhode Island."

176 **a sense of crisis emerged** In a statement, a Rhode Island Hospital spokeswoman wrote: "I would not describe the atmosphere as being one of crisis—it was more accurately one of demoralization among many. Many people felt beleaguered."

177 **to make sure time-outs occurred** The cameras were installed as part of a consent order with the state's department of health.

177 **A computerized system** Rhode Island Hospital Surgical Safety Backgrounder, provided by hospital administrators. More information on Rhode Island Hospital's safety initiatives is available at http://rhodeislandhospital. org/rih/quality/.

177 **But once a sense of crisis gripped** For more on how crises can create an atmosphere where change is possible in medicine, and how wrong-site surgeries occur, see Douglas McCarthy and David Blumenthal, "Stories from the Sharp End: Case Studies in Safety Improvement," *Milbank Quarterly* 84 (2006): 165–200; J. W. Senders et al., "The Egocentric Surgeon or the Roots of Wrong Side Surgery," *Quality and Safety in Health Care* 17 (2008): 396–400; Mary R. Kwaan et al., "Incidence, Patterns, and Prevention of Wrong-Site Surgery," *Archives of Surgery* 141, no. 4 (April 2006): 353–57.

177 **Other hospitals have made similar** For a discussion on this topic, see McCarthy and Blumenthal, "Stories from the Sharp End"; Atul Gawande, *Better: A Surgeon's Notes on Performance* (New York: Metropolitan Books, 2008); Atul Gawande, *The Checklist Manifesto: How to Get Things Right* (New York: Metropolitan Books, 2009).

178 **In the wake of that tragedy** NASA, "Report to the President: Actions to Implement the Recommendations of the Presidential Commission on the Space Shuttle *Challenger* Accident," July 14, 1986; Matthew W. Seeger, "The *Challenger* Tragedy and Search for Legitimacy," *Communication Studies* 37, no. 3 (1986): 147–57; John Noble Wilford, "New NASA System Aims to Encourage Blowing the Whistle," *The New York Times*, June 5, 1987; Joseph Lorenzo Hall, "*Columbia* and *Challenger*: Organizational Failure at NASA," Space Policy 19, no. 4 (November 2003), 239–47; Barbara Romzek and Melvin Dubnick, "Accountability in the Public Sector: Lessons from the *Challenger* Tragedy," *Public Administration Review* 47, no. 3 (May–June 1987): 227–38.

178 Then, a runway error Karl E. Weick, "The Vulnerable System: An Analysis of the Tenerife Air Disaster," *Journal of Management* 16, no. 3 (1990): 571–93; William Evan and Mark Manion, *Minding the Machines: Preventing Technological Disasters* (Upper Saddle River, N.J.: Prentice Hall Professional, 2002); Raimo P. Hämäläinen and Esa Saarinen, *Systems Intelligence: Discovering a Hidden Competence in Human Action and Organizational Life* (Helsinki: Helsinki University of Technology, 2004).

CHAPTER SEVEN

186 grab an extra box The details on subconscious tactics retailers use comes from Jeremy Caplan, "Supermarket Science," *Time,* May 24, 2007; Paco Underhill, *Why We Buy: The Science of Shopping* (New York: Simon and Schuster, 2000); Jack Hitt; "The Theory of Supermarkets," *The New York Times,* March 10, 1996; "The Science of Shopping: The Way the Brain Buys," *The Economist,* December 20, 2008; "Understanding the Science of Shopping," *Talk of the Nation,* National Public Radio, December 12, 2008; Malcolm Gladwell, "The Science of Shopping," *The New Yorker,* November 4, 1996.

186 to buy almost anything There are literally thousands of studies that have scrutinized how habits influence consumer behaviors—and how unconscious and semi-conscious urges influence decisions that might otherwise seem immune from habitual triggers. For more on these fascinating topics, see H. Aarts, A. van Knippenberg, and B. Verplanken, "Habit and Information Use in Travel Mode Choices," *Acta Psychologica* 96, nos. 1–2 (1997): 1–14; J. A. Bargh, "The Four Horsemen of Automaticity: Awareness, Efficiency, Intention, and Control in Social Cognition," in *Handbook of Social Cognition,* ed. R. S. Wyer, Jr., and T. K. Srull (Hillsdale, N.J.: Lawrence Erlbaum Associates, 1994); D. Bell, T. Ho, and C. Tang, "Determining Where to Shop: Fixed and Variable Costs of Shopping," *Journal of Marketing Research* 35, no. 3 (1998): 352–69; T. Betsch, S. Haberstroh, B. Molter, A. Glöckner, "Oops, I Did It Again—Relapse Errors in Routinized Decision Making," *Organizational Behavior and Human Decision Processes* 93, no. 1 (2004): 62–74; M. Cunha, C. Janiszewski, Jr., and J. Laran, "Protection of Prior Learning in Complex Consumer Learning Environments," *Journal of Consumer Research* 34, no. 6 (2008): 850–64; H. Aarts, U. Danner, and N. de Vries, "Habit Formation and Multiple Means to Goal Attainment: Repeated Retrieval of Target Means Causes Inhibited Access to Competitors," *Personality and Social Psychology Bulletin* 33, no. 10 (2007): 1367–79; E. Ferguson and P. Bibby, "Predicting Future Blood Donor Returns: Past Behavior, Intentions, and Observer Effects," *Health Psychology* 21, no. 5 (2002): 513–18; Edward Fox and John Semple, "Understanding 'Cherry Pickers': How Retail Customers Split Their Shopping Baskets," unpublished manuscript, Southern Methodist University, 2002; S. Gopinath, R. Blattberg, and E. Malthouse, "Are Revived Customers as Good as New?" unpublished manuscript, Northwestern University, 2002; H. Aarts, R. Holland, and D. Langendam, "Breaking and Creating Habits on the Working Floor: A Field-Experiment on the Power of Implementation Intentions," *Journal of Experimental Social Psychology* 42, no. 6 (2006): 776–83; Mindy Ji and Wendy Wood, "Purchase and Consumption Habits: Not Necessarily What You Intend," *Journal of Consumer Psychology* 17, no. 4 (2007): 261–76; S. Bellman, E. J. Johnson, and G. Lohse, "Cognitive Lock-In and the Power Law of Practice," *Journal of Marketing* 67, no. 2 (2003): 62–75; J. Bettman et al., "Adapting to Time Constraints," in *Time Pressure and Stressing Human Judgment and Decision Making*, ed. O. Svenson and J. Maule (New York: Springer, 1993); Adwait Khare and J. Inman, "Habitual Behavior in American Eating Patterns: The Role of Meal Occasions," *Journal of Consumer Research* 32, no. 4 (2006): 567–75; David Bell and R. Lal, "The Impact of Frequent Shopper Programs in Grocery Retailing," *Quantitative Marketing and Economics* 1, no. 2 (2002): 179–202; Yuping Liu, "The Long-Term Impact of Loyalty Programs on Consumer Purchase Behavior and Loyalty," *Journal of Marketing* 71, no. 4 (2007): 19–35; Neale Martin, *Habit: The 95% of Behavior Marketers Ignore* (Upper Saddle River, N.J.: FT Press, 2008); H. Aarts, K. Fujia, and K. C. McCulloch, "Inhibition in Goal Systems: A Retrieval-Induced Forgetting Account," *Journal of Experimental Social Psychology* 44, no. 3 (2008): 614–23; Gerald Häubl and K. B. Murray, "Explaining Cognitive Lock-In: The Role of Skill-Based Habits of Use in Consumer Choice," *Journal of Consumer Research* 34 (2007) 77–88; D. Neale, J. Quinn, and W. Wood, "Habits: A Repeat Performance," *Current Directions in Psychological Science* 15, no. 4 (2006) 198–202; R. L. Oliver, "Whence Consumer Loyalty?" *Journal of Marketing* 63 (1999): 33–44; C. T. Orleans, "Promoting the Maintenance of Health Behavior Change: Recommendations for the Next Generation of Research and Practice," *Health Psychology* 19 (2000): 76–83; Andy Ouellette and Wendy Wood, "Habit and Intention in Everyday Life: The Multiple Processes by Which Past Behavior Predicts Future Behavior," *Psychological Bulletin* 124, no. 1 (1998) 54–74; E. Iyer, D. Smith, and C. Park, "The Effects of Situational Factors on In-Store Grocery Shopping Behavior: The Role of Store Environment and Time Available for Shopping," *Journal of Consumer Research* 15, no. 4 (1989): 422–33; O. Amir, R. Dhar, and A. Pocheptsova, "Deciding Without Resources: Resource Depletion and Choice in Context," *Journal of Marketing Research* 46, no. 3 (2009): 344–55; H. Aarts, R. Custers, and P. Sheeran, "The Goal-Dependent Automaticity of Drinking Habits," *British Journal of Social Psychology* 44, no. 1 (2005): 47–63; S. Orbell and P.

Sheeran, "Implementation Intentions and Repeated Behavior: Augmenting the Predictive Validity of the Theory of Planned Behavior," *European Journal of Social Psychology* 29, nos. 2–3 (1999): 349–69; P. Sheeran, P. Gollwitzer, and P. Webb, "The Interplay Between Goal Intentions and Implementation Intentions," *Personality and Social Psychology Bulletin* 31, no. 1 (2005): 87–98; H. Shen and R. S. Wyer, "Procedural Priming and Consumer Judgments: Effects on the Impact of Positively and Negatively Valenced Information," *Journal of Consumer Research* 34, no. 5 (2007): 727–37; Itamar Simonson, "The Effect of Purchase Quantity and Timing on Variety-Seeking Behavior," *Journal of Marketing Research* 27, no. 2 (1990): 150–62; G. Taylor and S. Neslin, "The Current and Future Sales Impact of a Retail Frequency Reward Program," *Journal of Retailing* 81, no. 4, 293–305; H. Aarts and B. Verplanken, "Habit, Attitude, and Planned Behavior: Is Habit an Empty Construct or an Interesting Case of Goal-Directed Automaticity?" *European Review of Social Psychology* 10 (1999): 101–34; B. Verplanken, Henk Aarts, and Ad Van Knippenberg, "Habit, Information Acquisition, and the Process of Making Travel Mode Choices," *European Journal of Social Psychology* 27, no. 5 (1997): 539–60; B. Verplanken et al., "Attitude Versus General Habit: Antecedents of Travel Mode Choice," *Journal of Applied Social Psychology* 24, no. 4 (1994): 285–300; B. Verplanken et al., "Consumer Style and Health: The Role of Impulsive Buying in Unhealthy Eating," *Psychology and Health* 20, no. 4 (2005): 429–41; B. Verplanken et al., "Context Change and Travel Mode Choice: Combining the Habit Discontinuity and Self-Activation Hypotheses," *Journal of Environmental Psychology* 28 (2008): 121–27; Bas Verplanken and Wendy Wood, "Interventions to Break and Create Consumer Habits," *Journal of Public Policy and Marketing* 25, no. 1 (2006): 90–103; H. Evanschitzky, B. Ramaseshan, and V. Vogel, "Customer Equity Drivers and Future Sales," *Journal of Marketing* 72 (2008): 98–108; P. Sheeran and T. L. Webb, "Does Changing Behavioral Intentions Engender Behavioral Change? A Meta-Analysis of the Experimental Evidence," *Psychological Bulletin* 132, no. 2 (2006): 249–68; P. Sheeran, T. L. Webb, and A. Luszczynska, "Planning to Break Unwanted Habits: Habit Strength Moderates Implementation Intention Effects on Behavior Change," *British Journal of Social Psychology* 48, no. 3 (2009): 507–23; D. Wegner and R. Wenzlaff, "Thought Suppression," *Annual Review of Psychology* 51 (2000): 59–91; L. Lwin, A. Mattila, and J. Wirtz, "How Effective Are Loyalty Reward Programs in Driving Share of Wallet?" *Journal of Service Research* 9, no. 4 (2007): 327–34; D. Kashy, J. Quinn, and W. Wood, "Habits in Everyday Life: Thought, Emotion, and Action," *Journal of Personality and Social Psychology* 83, no. 6 (2002): 1281–97; L. Tam, M. Witt, and W. Wood (2005), "Changing Circumstances, Disrupting Habits," *Journal of Personality and Social Psychology* 88, no. 6 (2005): 918–33; Alison Jing Xu and Robert S. Wyer, "The Effect of Mind-sets on Consumer Decision Strategies," *Journal of Con-*

sumer Research 34, no. 4 (2007): 556–66; C. Cole, M. Lee, and C. Yoon, "Consumer Decision Making and Aging: Current Knowledge and Future Directions," *Journal of Consumer Psychology* 19 (2009): 2–16; S. Dhar, A. Krishna, and Z. Zhang, "The Optimal Choice of Promotional Vehicles: Front-Loaded or Rear-Loaded Incentives?" *Management Science* 46, no. 3 (2000): 348–62.

186 **"potato chips are on sale!"** C. Park, E. Iyer, and D. Smith, "The Effects of Situational Factors on In-Store Grocery Shopping Behavior: The Role of Store Environment and Time Available for Shopping," *The Journal of Consumer Research* 15, no. 4 (1989): 422–33. For more on this topic, see J. Belyavsky Bayuk, C. Janiszewski, and R. Leboeuf, "Letting Good Opportunities Pass Us By: Examining the Role of Mind-set During Goal Pursuit," *Journal of Consumer Research* 37, no. 4 (2010): 570–83; Ab Litt and Zakary L. Tormala, "Fragile Enhancement of Attitudes and Intentions Following Difficult Decisions," *Journal of Consumer Research* 37, no. 4 (2010): 584–98.

187 **University of Southern California** D. Neal and W. Wood, "The Habitual Consumer," *Journal of Consumer Psychology* 19, no. 4 (2009): 579–92. For more on similar research, see R. Fazio and M. Zanna, "Direct Experience and Attitude–Behavior Consistency," in *Advances in Experimental Social Psychology*, ed. L. Berkowitz (New York: Academic Press, 2005); R. Abelson and R. Schank, "Knowledge and Memory: The Real Story," in *Knowledge and Memory: The Real Story*, ed. R. S. Wyer, Jr. (Hillsdale, N.J.: Lawrence Erlbaum, 2004); Nobert Schwarz, "Meta-Cognitive Experiences in Consumer Judgment and Decision Making," *Journal of Consumer Psychology* 14, no. 4 (September 2004): 332–48; R. Wyer and A. Xu, "The Role of Behavioral Mindsets in Goal-Directed Activity: Conceptual Underpinnings and Empirical Evidence," *Journal of Consumer Psychology* 20, no. 2 (2010): 107–25.

188 **news or deals on cigarettes** Julia Angwin and Steve Stecklow, "'Scrapers' Dig Deep for Data on Web," *The Wall Street Journal*, October 12, 2010; Mark Maremont and Leslie Scism, "Insurers Test Data Profiles to Identify Risky Clients," *The Wall Street Journal*, November 19, 2010; Paul Sonne and Steve Stecklow, "Shunned Profiling Technology on the Verge of Comeback," *The Wall Street Journal*, November 24, 2010.

189 **Pole flashed a slide** This slide is from a keynote speech by Pole at Predicted Analytics World, New York, October 20, 2009. It is no longer available online. Additionally, see Andrew Pole, "Challenges of Incremental Sales Modeling in Direct Marketing."

192 **buying different brands of beer** It's difficult to make specific correlations between types of life changes and specific products. So, while we know that people who move or get divorced will change their buying patterns, we don't know that divorce always influences beer, or that a new home always influences cereal purchases. But the general trend holds. Alan Andreasen, "Life Status Changes and Changes in Consumer Preferences and Satisfaction," *Journal of Consumer Research* 11, no. 3 (1984): 784–94. For more on this topic, see E. Lee, A. Mathur, and G. Moschis, "A Longitudinal Study of the Effects of Life Status Changes on Changes in Consumer Preferences," *Journal of the Academy of Marketing Science* 36, no. 2 (2007): 234–46; L. Euehun, A. Mathur, and G. Moschis, "Life Events and Brand Preferences Changes," *Journal of Consumer Behavior* 3, no. 2 (2003): 129–41.

192   and they care quite a bit For more on the fascinating topic of how particular moments offer opportunities for marketers (or government agencies, health activists, or anyone else, for that matter) to influence habits, see Bas Verplanken and Wendy Wood, "Interventions to Break and Create Consumer Habits," *Journal of Public Policy and Marketing* 25, no. 1 (2006): 90–103; D. Albarracin, A. Earl, and J. C. Gillette, "A Test of Major Assumptions About Behavior Change: A Comprehensive Look at the Effects of Passive and Active HIV-Prevention Interventions Since the Beginning of the Epidemic," *Psychological Bulletin* 131, no. 6 (2005): 856–97; T. Betsch, J. Brinkmann, and K. Fiedler, "Behavioral Routines in Decision Making: The Effects of Novelty in Task Presentation and Time Pressure on Routine Maintenance and Deviation," *European Journal of Social Psychology* 28, no. 6 (1998): 861–78; L. Breslow, "Social Ecological Strategies for Promoting Healthy Lifestyles," *American Journal of Health Promotion* 10, no. 4 (1996), 253–57; H. Buddelmeyer and R. Wilkins, "The Effects of Smoking Ban Regulations on Individual Smoking Rates," Melbourne Institute Working Paper Series no. 1737, Melbourne Institute of Applied Economic and Social Research, University of Melbourne, 2005; P. Butterfield, "Thinking Upstream: Nurturing a Conceptual Understanding of the Societal Context of Health Behavior," *Advances in Nursing Science* 12, no. 2 (1990): 1–8; J. Derzon and M. Lipsey, "A Meta-Analysis of the Effectiveness of Mass Communication for Changing Substance-Use Knowledge, Attitudes, and Behavior," in *Mass Media and Drug Prevention: Classic and Contemporary Theories and Research*, ed. W. D. Crano and M. Burgoon (East Sussex, U.K.: Psychology, 2001); R. Fazio, J. Ledbetter, and T. Ledbetter, "On the Costs of Accessible Attitudes: Detecting That the Attitude Object Has Changed," *Journal of Personality and Social Psychology* 78, no. 2 (2000): 197–210; S. Fox et al., "Competitive Food Initiatives in Schools and Overweight in Children: A Review of the Evidence," *Wisconsin Medical Journal* 104, no. 8 (2005): 38–43; S. Fujii, T. Gärling, and R. Kitamura, "Changes in Drivers' Perceptions and Use of Public Transport During a Freeway Closure: Effects of Temporary Structural Change on Cooperation in a Real-Life Social Dilemma," *Environment and Behavior* 33, no. 6 (2001): 796–808; T. Heatherton and P. Nichols, "Personal Accounts of Successful Versus Failed Attempts at Life Change," *Personality and Social Psy-*

chology Bulletin 20, no. 6 (1994): 664–75; J. Hill and H. R. Wyatt, "Obesity and the Environment: Where Do We Go from Here?" *Science* 299, no. 5608 (2003): 853–55; P. Johnson, R. Kane, and R. Town, "A Structured Review of the Effect of Economic Incentives on Consumers' Preventive Behavior," *American Journal of Preventive Medicine* 27, no. 4 (2004): 327–52; J. Fulkerson, M. Kubrik, and L. Lytle, "Fruits, Vegetables, and Football: Findings from Focus Groups with Alternative High School Students Regarding Eating and Physical Activity," *Journal of Adolescent Health* 36, no. 6 (2005): 494–500; M. Abraham, S. Kalmenson, and L. Lodish, "How T.V. Advertising Works: A Meta-Analysis of 389 Real World Split Cable T.V. Advertising Experiments," *Journal of Marketing Research* 32, no. 5 (1995): 125–39; J. McKinlay, "A Case for Re-Focusing Upstream: The Political Economy of Illness," in *Applying Behavioral Science to Cardiovascular Risk*, ed. A. J. Enelow and J. B. Henderson (New York: American Heart Association, 1975); N. Milio, "A Framework for Prevention: Changing Health-Damaging to Health-Generating Life Patterns," *American Journal of Public Health* 66, no. 5 (1976): 435–39; S. Orbell, "Intention-Behavior Relations: A Self-Regulatory Perspective," in *Contemporary Perspectives on the Psychology of Attitudes*, ed. G. Haddock and G. Maio (New York: Psychology Press, 2004); C. T. Orleans, "Promoting the Maintenance of Health Behavior Change: Recommendations for the Next Generation of Research and Practice," *Health Psychology* 19, no. 1 (2000): 76–83; C. G. DiClemente, J. C. Norcross, and J. Prochaska, "In Search of How People Change: Applications to Addictive Behaviors," *American Psychologist* 47, no. 9 (1992): 1102–14; J. Quinn and W. Wood, "Inhibiting Habits and Temptations: Depends on Motivational Orientation," 2006 manuscript under editorial review; T. Mainieri, S. Oskamp, and P. Schultz, "Who Recycles and When? A Review of Personal and Structural Factors," *Journal of Environmental Psychology* 15, no. 2 (1995): 105–21; C. D. Jenkins, C. T. Orleans, and T. W. Smith, "Prevention and Health Promotion: Decades of Progress, New Challenges, and an Emerging Agenda," *Health Psychology* 23, no. 2 (2004): 126–31; H. C. Triandis, "Values, Attitudes, and Interpersonal Behavior," *Nebraska Symposium on Motivation* 27 (1980): 195–259.

192 before a child's first birthday "Parents Spend £5,000 on Newborn Baby Before Its First Birthday," *Daily Mail*, September 20, 2010.

193 $36.3 billion a year Brooks Barnes, "Disney Looking into Cradle for Customers," *The New York Times*, February 6, 2011.

195 Jenny Ward, a twenty-three-year-old The names in this paragraph are pseudonyms, used to illustrate the types of customers Target's models can detect. These are not real shoppers.

196 profile their buying habits "McDonald's, CBS, Mazda, and Microsoft Sued for 'History Sniffing,'" Forbes.com, January 3, 2011.

196 ferret out their mailing addresses Terry Baynes, "California Ruling Sets Off More Credit Card Suits," Reuters, February 16, 2011.

198 forecasted if a tune was likely to succeed A. Elberse, J. Eliashbert, and J. Villanueva, "Polyphonic HMI: Mixing Music with Math," *Harvard Business Review,* August 24, 2005.

198 thirty-seven times throughout the month My thanks to Adam Foster, director of data services, Nielsen BDS.

199 Listeners didn't just dislike "Hey Ya!" My thanks to Paul Heine, now of *Inside Radio;* Paul Heine, "Fine-tuning People Meter," *Billboard,* November 6, 2004; Paul Heine, "Mscore Data Shows Varying Relationship with Airplay," *Billboard,* April 3, 2010.

199 make "Hey Ya!" into a hit In fact-checking communications, Steve Bartels, the Arista promotions executive, emphasized that he saw the fact that "Hey Ya!" was polarizing as a good thing. The song was released and promoted with another tune—"The Way You Move"—that was the other big single from OutKast's two-disc release *Speakerboxxx/The Love Below.* "You want there to be a reaction," Bartels told me. "Some of the smarter [program directors] looked at the polarization as an opportunity to give the station an identity. The fact that there was a quick turn-off reaction, to me, doesn't mean we're not succeeding. It's my job to convince PDs that's why they should look at this song."

201 they stayed glued Stephanie Clifford, "You Never Listen to Celine Dion? Radio Meter Begs to Differ," *The New York Times,* December 15, 2009; Tim Feran, "Why Radio's Changing Its Tune," *The Columbus Dispatch,* June 13, 2010.

202 the superior parietal cortex G. S. Berns, C. M. Capra, and S. Moore, "Neural Mechanisms of the Influence of Popularity on Adolescent Ratings of Music," *NeuroImage* 49, no. 3 (2010): 2687–96; J. Bharucha, F. Musiek, and M. Tramo, "Music Perception and Cognition Following Bilateral Lesions of Auditory Cortex," *Journal of Cognitive Neuroscience* 2, no. 3 (1990): 195–212; Stefan Koelsch and Walter Siebel, "Towards a Neural Basis of Music Perception," *Trends in Cognitive Sciences* 9, no. 12 (2005): 578–84; S. Brown, M. Martinez, and L. Parsons, "Passive Music Listening Spontaneously Engages Limbic and Paralimbic Systems," *NeuroReport* 15, no. 13 (2004): 2033–37; Josef Rauschecker, "Cortical Processing of Complex Sounds," *Current Opinion in Neurobiology* 8, no. 4 (1998): 516–21; J. Kaas, T. Hackett, and M. Tramo, "Auditory Processing in Primate Cerebral Cortex," *Current Opinion in Neurobiology* 9, no. 2 (1999): 164–70; S. Koelsch, "Neural Substrates of Processing Syntax and Semantics in Music," *Current Opinion in Neurobiology* 15 (2005): 207–12; A. Lahav, E. Saltzman, and G. Schlaug, "Action Representation of Sound: Audiomotor Recognition Network While Listening to Newly Acquired Actions,"

Journal of Neuroscience 27, no. 2 (2007): 308–14; D. Levitin and V. Menon, "Musical Structure Is Processed in 'Language' Areas of the Brain: A Possible Role for Brodmann Area 47 in Temporal Coherence," *NeuroImage* 20, no. 4 (2003): 2142–52; J. Chen, V. Penhume, and R. Zatorre, "When the Brain Plays Music: Auditory-Motor Interactions in Music Perception and Production," *Nature Reviews Neuroscience* 8, 547–58.

202 **a cacophony of noise** N. S. Rickard and D. Ritossa, "The Relative Utility of 'Pleasantness' and 'Liking' Dimensions in Predicting the Emotions Expressed by Music," *Psychology of Music* 32, no. 1 (2004): 5–22; G. Berns, C. Capra, and S. Moore, "Neural Mechanisms of the Influence of Popularity on Adolescent Ratings of Music," *NeuroImage* 49, no. 3 (2010): 2687–96; David Hargreaves and Adrian North, "Subjective Complexity, Familiarity, and Liking for Popular Music," *Psychomusicology* 14, no. 1996 (1995): 77–93. For more on this fascinating topic of how familiarity influences attractiveness across numerous senses, see also G. Berns, S. McClure, and G. Pagnoni, "Predictability Modulates Human Brain Response to Reward," *Journal of Neuroscience* 21, no. 8 (2001): 2793–98; D. Brainard, "The Psychophysics Toolbox," *Spatial Vision* 10 (1997): 433–36; J. Cloutier, T. Heatherton, and P. Whalen, "Are Attractive People Rewarding? Sex Differences in the Neural Substrates of Facial Attractiveness," *Journal of Cognitive Neuroscience* 20, no. 6 (2008): 941–51; J. Kable and P. Glimcher, "The Neural Correlates of Subjective Value During Intertemporal Choice," *Nature Neuroscience* 10, no. 12 (2007): 1625–33; S. McClure et al., "Neural Correlates of Behavioral Preference for Culturally Familiar Drinks," *Neuron* 44, no. 2 (2004): 379–87; C. J. Assad and Padoa-Schioppa, "Neurons in the Orbitofrontal Cortex Encode Economic Value," *Nature* 441, no. 7090 (2006): 223–26; H. Plassmann et al., "Marketing Actions Can Modulate Neural Representations of Experienced Pleasantness," *Proceedings of the National Academy of Science* 105, no. 3 (2008): 1050–54; Muzafer Sherif, *The Psychology of Social Norms* (New York: Harper and Row, 1936); Wendy Wood, "Attitude Change: Persuasion and Social Influence," *Annual Review of Psychology* 51 (2000): 539–70; Gustave Le Bon, *The Crowd: A Study of the Popular Mind* (Mineola, N.Y.: Dover Publications, 2001); G. Berns et al., "Neural Mechanisms of Social Influence in Consumer Decisions," working paper, 2009; G. Berns et al., "Nonlinear Neurobiological Probability Weighting Functions for Aversive Outcomes," *NeuroImage* 39, no. 4 (2008): 2047–57; G. Berns et al., "Neurobiological Substrates of Dread," *Science* 312, no. 5 (2006): 754–58; G. Berns, J. Chappelow, and C. Zink, "Neurobiological Correlates of Social Conformity and Independence During Mental Rotation," *Biological Psychiatry* 58, no. 3 (2005): 245–53; R. Bettman, M. Luce, and J. Payne, "Constructive Consumer Choice Processes," *Journal of Consumer Research* 25, no. 3 (1998): 187–217; A. Blood and R. Zatorre, "Intensely Pleasurable Responses to Music

Correlate with Activity in Brain Regions Implicated in Reward and Emotion," *Proceedings of the National Academy of Science* 98, no. 20 (2001): 11818–23; C. Camerer, G. Loewenstein, and D. Prelec, "Neuroeconomics: How Neuroscience Can Inform Economics," *Journal of Economic Literature* 43, no. 1 (2005): 9–64; C. Capra et al., "Neurobiological Regret and Rejoice Functions for Aversive Outcomes," *NeuroImage* 39, no. 3 (2008): 1472–84; H. Critchley et al., "Neural Systems Supporting Interoceptive Awareness," *Nature Neuroscience* 7, no. 2 (2004): 189–95; H. Bayer, M. Dorris, and P. Glimcher, "Physiological Utility Theory and the Neuroeconomics of Choice," *Games and Economic Behavior* 52, no. 2, 213–56; M. Brett and J. Grahn, "Rhythm and Beat Perception in Motor Areas of the Brain," *Journal of Cognitive Neuroscience* 19, no. 5 (2007): 893–906; A. Hampton and J. O'Doherty, "Decoding the Neural Substrates of Reward-Related Decision-Making with Functional MRI," *Proceedings of the National Academy of Science* 104, no. 4 (2007): 1377–82; J. Birk et al., "The Cortical Topography of Tonal Structures Underlying Western Music," *Science* 298 (2002): 2167–70; B. Knutson et al., "Neural Predictors of Purchases," *Neuron* 53, no. 1 (2007): 147–56; B. Knutson et al., "Distributed Neural Representation of Expected Value," *Journal of Neuroscience* 25, no. 19 (2005): 4806–12; S. Koelsch, "Neural Substrates of Processing Syntax and Semantics in Music," *Current Opinion in Neurobiology* 15, no. 2 (2005): 207–12; T. Fritz et al., "Adults and Children Processing Music: An fMRI Study," *NeuroImage* 25 (2005): 1068–76; T. Fritz et al., "Investigating Emotion with Music: An fMRI Study," *Human Brain Mapping* 27 (2006): 239–50; T. Koyama et al., "The Subjective Experience of Pain: Where Expectations Becomes Reality," *Proceedings of the National Academy of Science* 102, no. 36 (2005): 12950–55; A. Lahav, E. Saltzman, and G. Schlaug, "Action Representation of Sound: Audiomotor Recognition Network While Listening to Newly Acquired Actions," *Journal of Neuroscience* 27, no. 2 (2007): 308–14; D. Levitin and V. Menon, "Musical Structure Is Processed in 'Language' Areas of the Brain: A Possible Role for Brodmann Area 47 in Temporal Coherence," *NeuroImage* 20, no. 4 (2003): 2142–52; G. Berns and P. Montague, "Neural Economics and the Biological Substrates of Valuation," *Neuron* 36 (2002): 265–84; C. Camerer, P. Montague, and A. Rangel, "A Framework for Studying the Neurobiology of Value-Based Decision Making," *Nature Reviews Neuroscience* 9 (2008): 545–56; C. Chafe et al., "Neural Dynamics of Event Segmentation in Music: Converging Evidence for Dissociable Ventral and Dorsal Networks," *Neuron* 55, no. 3 (2007): 521–32; Damian Ritossa and Nikki Rickard, "The Relative Utility of 'Pleasantness' and 'Liking' Dimensions in Predicting the Emotions Expressed by Music," *Psychology of Music* 32, no. 1 (2004): 5–22; Gregory S. Berns et al., "Neural Mechanisms of the Influence of Popularity on Adolescent Ratings of Music," *NeuroImage* 49, no. 3 (2010): 2687–96; Adrian North and David Hargreaves, "Subjective Complexity, Familiarity, and Liking for Popular Music," *Psychomusicology* 14, nos. 1–2 (1995): 77–93; Walter Ritter, Elyse Sussman, and Herbert Vaughan, "An Investigation of the Auditory Streaming Effect Using Event-Related Brain Potentials," *Psychophysiology* 36, no. 1 (1999): 22–34; Elyse Sussman, Rika Takegata, and István Winkler, "Event-Related Brain Potentials Reveal Multiple Stages in the Perceptual Organization of Sound," *Cognitive Brain Research* 25, no. 1 (2005): 291–99; Isabelle Peretz and Robert Zatorre, "Brain Organization for Music Processing," *Annual Review of Psychology* 56, no. 1 (2005): 89–114.

204 a black market for poultry Charles Grutzner, "Horse Meat Consumption by New Yorkers Is Rising," *The New York Times*, September 25, 1946.

205 camouflage it in everyday garb It is worth noting that this was only one of the committee's many findings (which ranged far and wide). For a fascinating study on the committee and its impacts, see Brian Wansink, "Changing Eating Habits on the Home Front: Lost Lessons from World War II Research," *Journal of Public Policy and Marketing* 21, no. 1 (2002): 90–99.

205 present-day researcher Wansink, "Changing Eating Habits on the Home Front."

206 cheer for steak and kidney pie" Brian Wansink, *Marketing Nutrition: Soy, Functional Foods, Biotechnology, and Obesity* (Champaign: University of Illinois, 2007).

206 it was up 50 percent Dan Usher, "Measuring Real Consumption from Quantity Data, Canada 1935–1968," in *Household Production and Consumption*, ed. Nestor Terleckyj (New York: National Bureau of Economic Research, 1976). It's very hard to get U.S. data on offal consumption, and so these calculations are based on Canadian trends, where data on the topic is more plentiful. In interviews, U.S. officials said that Canada is a fair proxy for U.S. trends. The calculations in Usher's paper draw on calculations of "canned meat," which contained offal.

210 "sizable increases in trips and sales" Target Corporation Analyst Meeting, October 18, 2005.

CHAPTER EIGHT

215 a ten-cent fare into the till For my understanding of the Montgomery bus boycott, I am indebted to those historians who have made themselves available to me, including John A. Kirk and Taylor Branch. My understanding of these events also draws on John A. Kirk, *Martin Luther King, Jr.: Profiles in Power* (New York: Longman, 2004); Taylor Branch, *Parting the Waters: America in the King Years, 1954–63* (New York: Simon and Schuster, 1988); Taylor Branch, *Pillar of Fire: America in the King Years, 1963–65* (New York:

Simon and Schuster, 1998); Taylor Branch, *At Canaan's Edge: America in the King Years, 1965–68* (New York: Simon and Schuster, 2006); Douglas Brinkley, *Mine Eyes Have Seen the Glory: The Life of Rosa Parks* (London: Weidenfeld and Nicolson, 2000); Martin Luther King, Jr., *Stride Toward Freedom: The Montgomery Story* (New York: Harper and Brothers, 1958); Clayborne Carson, ed., *The Papers of Martin Luther King, Jr.*, vol. 1, *Called to Serve* (Berkeley: University of California, 1992), vol. 2, *Rediscovering Precious Values* (1994), vol. 3, *Birth of a New Age* (1997), vol. 4, *Symbol of the Movement* (2000), vol. 5, *Threshold of a New Decade* (2005); Aldon D. Morris, *The Origins of the Civil Rights Movement* (New York: Free Press, 1986); James Forman, *The Making of Black Revolutionaries* (Seattle: University of Washington, 1997). Where not cited, facts draw primarily from those sources.

216 **"You may do that," Parks said** Henry Hampton and Steve Fayer, eds., *Voices of Freedom: An Oral History of the Civil Rights Movement from the 1950s Through the 1980s* (New York: Bantam Books, 1995); Rosa Parks, *Rosa Parks: My Story* (New York: Puffin, 1999).

216 **"the law is the law"** John A. Kirk, *Martin Luther King, Jr.: Profiles in Power* (New York: Longman, 2004).

217 **a three-part process** For more on the sociology of movements, see G. Davis, D. McAdam, and W. Scott, *Social Movements and Organizations* (New York: Cambridge University, 2005); Robert Crain and Rita Mahard, "The Consequences of Controversy Accompanying Institutional Change: The Case of School Desegregation," *American Sociological Review* 47, no. 6 (1982): 697–708; Azza Salama Layton, "International Pressure and the U.S. Government's Response to Little Rock," *Arkansas Historical Quarterly* 56, no. 3 (1997): 257–72; Brendan Nelligan, "The Albany Movement and the Limits of Nonviolent Protest in Albany, Georgia, 1961–1962," Providence College Honors Thesis, 2009; Charles Tilly, *Social Movements, 1768–2004* (London: Paradigm, 2004); Andrew Walder, "Political Sociology and Social Movements," *Annual Review of Sociology* 35 (2009): 393–412; Paul Almeida, *Waves of Protest: Popular Struggle in El Salvador, 1925–2005* (Minneapolis: University of Minnesota, 2008); Robert Benford, "An Insider's Critique of the Social Movement Framing Perspective," *Sociological Inquiry* 67, no. 4 (1997): 409–30; Robert Benford and David Snow, "Framing Processes and Social Movements: An Overview and Assessment," *Annual Review of Sociology* 26 (2000): 611–39; Michael Burawoy, *Manufacturing Consent: Changes in the Labor Process Under Monopoly Capitalism* (Chicago: University of Chicago, 1979); Carol Conell and Kim Voss, "Formal Organization and the Fate of Social Movements: Craft Association and Class Alliance in the Knights of Labor," *American Sociological Review* 55, no. 2 (1990): 255–69; James Davies, "Toward a Theory of Revolution," *American Sociological Review* 27, no. 1 (1962): 5–18; William Gamson, *The Strategy of Social Protest* (Homewood,

Ill.: Dorsey, 1975); Robert Benford, "An Insider's Critique of the Social Move-
ment Framing Perspective," *Sociological Inquiry* 67, no. 4 (1997): 409–30; Jeff
Goodwin, *No Other Way Out: States and Revolutionary Movements, 1945–1991*
(New York: Cambridge University, 2001); Jeff Goodwin and James Jasper, eds.,
Rethinking Social Movements: Structure, Meaning, and Emotion (Lanham, Md.:
Rowman and Littlefield, 2003); Roger Gould, "Multiple Networks and Mobilization in the Paris Commune, 1871," *American Sociological Review* 56, no. 6
(1991): 716–29; Joseph Gusfield, "Social Structure and Moral Reform: A Study
of the Woman's Christian Temperance Union," *American Journal of Sociology*
61, no. 3 (1955): 221–31; Doug McAdam, *Political Process and the Development
of Black Insurgency, 1930–1970* (Chicago: University of Chicago, 1982); Doug
McAdam, "Recruitment to High-Risk Activism: The Case of Freedom Summer," *American Journal of Sociology* 92, no. 1 (1986): 64–90; Doug McAdam,
"The Biographical Consequences of Activism," *American Sociological Review* 54,
no. 5 (1989): 744–60; Doug McAdam, "Conceptual Origins, Current Problems,
Future Directions," in *Comparative Perspectives on Social Movements: Political
Opportunities, Mobilizing Structures, and Cultural Framings*, ed. Doug McAdam,
John McCarthy, and Mayer Zald (New York: Cambridge University, 1996);
Doug McAdam and Ronnelle Paulsen, "Specifying the Relationship Between
Social Ties and Activism," *American Journal of Sociology* 99, no. 3 (1993): 640–
67; D. McAdam, S. Tarrow, and C. Tilly, *Dynamics of Contention* (Cambridge:
Cambridge University, 2001); Judith Stepan-Norris and Judith Zeitlin, "'Who
Gets the Bird?' or, How the Communists Won Power and Trust in America's
Unions," *American Sociological Review* 54, no. 4 (1989): 503–23; Charles Tilly,
From Mobilization to Revolution (Reading, Mass.: Addison-Wesley, 1978).

218   **talking back to a Montgomery bus driver** Phillip Hoose, *Claudette Colvin: Twice Toward Justice* (New York: Farrar, Straus and Giroux, 2009).

218   **and refusing to move** Ibid.

218   **sitting next to a white man** Russell Freedman, *Freedom Walkers: The
Story of the Montgomery Bus Boycott* (New York: Holiday House, 2009).

218   **"indignities which came with it"** Martin Luther King, Jr., *Stride Toward
Freedom* (New York: Harper and Brothers, 1958).

219   **"a dozen or so sociopaths"** Taylor Branch, *Parting the Waters: America in
the King Years, 1954–63* (New York: Simon and Schuster, 1988).

221   **"white folks will kill you"** Douglas Brinkley, *Mine Eyes Have Seen the
Glory: The Life of Rosa Parks* (London: Weidenfeld and Nicolson, 2000).

221   **"happy to go along with it"** John A. Kirk, *Martin Luther King, Jr.: Profiles
in Power* (New York: Longman, 2004).

221   **in protest of the arrest and trial** Carson, *Papers of Martin Luther King, Jr.*

223 how 282 men had found their Mark Granovetter, *Getting a Job: A Study of Contacts and Careers* (Chicago: University of Chicago, 1974).

224 we would otherwise never hear about Andreas Flache and Michael Macy, "The Weakness of Strong Ties: Collective Action Failure in a Highly Cohesive Group," *Journal of Mathematical Sociology* 21 (1996): 3–28. For more on this topic, see Robert Axelrod, *The Evolution of Cooperation* (New York: Basic Books, 1984); Robert Bush and Frederick Mosteller, *Stochastic Models for Learning* (New York: Wiley, 1984); I. Erev, Y. Bereby-Meyer, and A. E. Roth, "The Effect of Adding a Constant to All Payoffs: Experimental Investigation and Implications for Reinforcement Learning Models," *Journal of Economic Behavior and Organization* 39, no. 1 (1999): 111–28; A. Flache and R. Hegselmann, "Rational vs. Adaptive Egoism in Support Networks: How Different Micro Foundations Shape Different Macro Hypotheses," in *Game Theory, Experience, Rationality: Foundations of Social Sciences, Economics, and Ethics in Honor of John C. Harsanyi (Yearbook of the Institute Vienna Circle)*, ed. W. Leinfellner and E. Köhler (Boston: Kluwer, 1997), 261–75; A. Flache and R. Hegselmann, "Rationality vs. Learning in the Evolution of Solidarity Networks: A Theoretical Comparison," *Computational and Mathematical Organization Theory* 5, no. 2 (1999): 97–127; A. Flache and R. Hegselmann, "Dynamik Sozialer Dilemma-Situationen," final research report of the DFG-Project Dynamics of Social Dilemma Situations, University of Bayreuth, Department of Philosophie, 2000; A. Flache and Michael Macy, "Stochastic Collusion and the Power Law of Learning," *Journal of Conflict Resolution* 46, no. 5 (2002): 629–53; Michael Macy, "Learning to Cooperate: Stochastic and Tacit Collusion in Social Exchange," *American Journal of Sociology* 97, no. 3 (1991): 808–43; E. P. H. Zeggelink, "Evolving Friendship Networks: An Individual-Oriented Approach Implementing Similarity," *Social Networks* 17 (1996): 83–110; Judith Blau, "When Weak Ties Are Structured," unpublished manuscript, Department of Sociology, State University of New York, Albany, 1980; Peter Blau, "Parameters of Social Structure," *American Sociological Review* 39, no. 5 (1974): 615–35; Scott Boorman, "A Combinatorial Optimization Model for Transmission of Job Information Through Contact Networks," *Bell Journal of Economics* 6, no. 1 (1975): 216–49; Ronald Breiger and Philippa Pattison, "The Joint Role Structure of Two Communities' Elites," *Sociological Methods and Research* 7, no. 2 (1978): 213–26; Daryl Chubin, "The Conceptualization of Scientific Specialties," *Sociological Quarterly* 17, no. 4 (1976): 448–76; Harry Collins, "The TEA Set: Tacit Knowledge and Scientific Networks," *Science Studies* 4, no. 2 (1974): 165–86; Rose Coser, "The Complexity of Roles as Seedbed of Individual Autonomy," in *The Idea of Social Structure: Essays in Honor of Robert Merton*, ed. L. Coser (New York: Harcourt, 1975); John Delany, "Aspects of Donative Resource Allocation and the Efficiency of Social Networks: Simulation Models of Job Vacancy Information Transfers

Through Personal Contacts," PhD diss., Yale University, 1980; E. Ericksen and W. Yancey, "The Locus of Strong Ties," unpublished manuscript, Department of Sociology, Temple University, 1980.

224 most of the population will be untouched Mark Granovetter, "The Strength of Weak Ties: A Network Theory Revisited," *Sociological Theory* 1 (1983): 201–33.

226 registering black voters in the South McAdam, "Recruitment to High-Risk Activism."

226 more than three hundred of those invited Ibid.; Paulsen, "Specifying the Relationship Between Social Ties and Activism."

226 participated in Freedom Summer In a fact-checking email, McAdam provided a few details about the study's genesis: "My initial interest was in trying to understand the links between the civil rights movement and the other early new left movements, specifically the student movement, the anti-war movement, and women's liberation movement. It was only after I found the applications and realized that some were from volunteers and others from 'no shows' that I got interested in explaining (a) why some made it to Mississippi and others didn't, and (b) the longer term impact of going/not-going on the two groups."

229 impossible for them to withdraw In another fact-checking email, McAdam wrote: "For me the significance of the organizational ties is not that they make it 'impossible' for the volunteer to withdraw, but that they insure that the applicant will likely receive lots of support for the link between the salient identity in question (i.e., Christian) and participation in the summer project. As I noted in [an article] 'it is a strong subjective identification with a particular identity, *reinforced by organizational ties* that is especially likely to encourage participation.'"

230 "getting together there without you" Tom Mathews and Roy Wilkins, *Standing Fast: The Autobiography of Roy Wilkins* (Cambridge, Mass.: Da Capo, 1994).

230 "boycott of city buses Monday" Branch, *Parting the Waters*.

231 "singing out, 'No riders today'" King, *Stride Toward Freedom*; James M. Washington, *A Testament of Hope: The Essential Writings and Speeches of Martin Luther King, Jr.* (New York: HarperCollins, 1990).

232 was in doubt King, *Stride Toward Freedom*.

233 drawing circles around major U.S. cities For understanding Pastor Warren's story, I am indebted to Rick Warren, Glenn Kruen, Steve Gladen, Jeff Sheler, Anne Krumm, and the following books: Jeffrey Sheler, *Prophet of Purpose: The Life of Rick Warren* (New York: Doubleday, 2009); Rick Warren, *The Purpose-Driven Church* (Grand Rapids, Mich.: Zondervan, 1995); and the following articles: Barbara Bradley, "Marketing That New-Time Religion," *Los Angeles Times,* December 10, 1995; John Wilson, "Not Just Another Mega Church," *Christianity Today,* December 4, 2000; "Therapy of the Masses," *The Economist,* November 6, 2003; "The Glue of Society," *The Economist,* July 14, 2005; Malcolm Gladwell, "The Cellular Church," *The New Yorker,* September 12, 2005; Alex MacLeod, "Rick Warren: A Heart for the Poor," *Presbyterian Record,* January 1, 2008; Andrew, Ann, and John Kuzma, "How Religion Has Embraced Marketing and the Implications for Business," *Journal of Management and Marketing Research* 2 (2009): 1–10.

233 **"our destination was a settled issue"** Warren, *Purpose-Driven Church.*

234 **"any chance of liberating multitudes"** Donald McGavran, *The Bridges of God* (New York: Friendship Press, 1955). Italics added.

235 **"How to Survive Under Stress"** Sheler, *Prophet of Purpose.*

236 **"I'm going to have to sit down"** In a fact-checking email a Saddleback spokesperson, provided additional details: "Rick suffers from a brain chemistry disorder that makes him allergic to adrenaline. This genetic problem resists medication and makes public speaking painful, with blurred vision, headaches, hot flashes, and panic. Symptoms usually last around fifteen minutes; by that time, enough adrenaline is expended so the body can return to normal function. (His adrenaline rushes, like any speaker might experience, whenever he gets up to preach.) Pastor Rick says this weakness keeps him dependent on God."

238 **"habits that will help you grow"** *Discovering Spiritual Maturity,* Class 201, published by Saddleback Church, http://www.saddlebackresources.com/CLASS-201-Discovering-Spiritual-Maturity-Complete-Kit-Download-P3532.aspx.

239 **"we just . . . get out of your way"** In a fact-checking email a Saddleback spokesperson said that while an important tenet of Saddleback is teaching people to guide themselves, "this implies that each person can go in any direction they choose. Biblical principles/guidelines have a clear direction. The goal of small group study is to teach people the spiritual disciplines of faith *and* everyday habits that can be applied to daily life."

239 **"community to continue the struggle"** Martin Luther King, Jr., *The Autobiography of Martin Luther King, Jr.,* ed. Clayborne Carson (New York: Grand Central, 2001).

240 **"shall perish by the sword"** Carson; King,

243 **segregation law violated the Constitution** *Browder v. Gayle,* 352 U.S. 903 (1956).

243 **and sat in the front** Washington, *Testament of Hope*.

243 **"glad to have you"** Kirk, *Martin Luther King, Jr.*

243 **"work and worry of the boycott"** Ibid.

CHAPTER NINE

245 **reorganizing the silverware drawer** "Angie Bachmann" is a pseudonym. Reporting for her story is based on more than ten hours of interviews with Bachmann, additional interviews with people who know Bachmann, and dozens of news articles and court filings. However, when Bachmann was presented with fact-checking questions, she declined to participate except to state that almost all details were inaccurate—including those she had previously confirmed, as well as facts confirmed by other sources, in court records, or by public documents—and then she cut off communication.

247 **"while thousands are injured"** *The Writings of George Washington*, vol. 8, ed. Jared Sparks (1835).

248 **swelled by more than $269 million** Iowa Racing and Gaming Commission, Des Moines, Iowa, 2010.

251 **"What have I done?"** Simon de Bruxelles, "Sleepwalker Brian Thomas Admits Killing Wife While Fighting Intruders in Nightmare," *The Times*, November 18, 2009.

252 **"I thought somebody had broken in"** Jane Mathews, "My Horror, by Husband Who Strangled Wife in Nightmare," *Daily Express*, December 16, 2010.

252 **"She's my world"** Simon de Bruxelles, "Sleepwalker Brian Thomas Admits Killing Wife While Fighting Intruders in Nightmare." *The Times*, November 18, 2009.

254 **annoying but benign problem** In some instances, people sleepwalk while they experience dreams, a condition known as REM sleep behavior disorder (see C. H. Schenck et al., "Motor Dyscontrol in Narcolepsy: Rapid-Eye-Movement [REM] Sleep Without Atonia and REM Sleep Behavior Disorder," *Annals of Neurology* 32, no. 1 [July 1992]: 3–10). In other instances, people are not dreaming, but move nonetheless.

254 **something called *sleep terrors*** C. Bassetti, F. Siclari, and R. Urbaniok, "Violence in Sleep," *Schweizer Archiv Fur Neurologie und Psychiatrie* 160, no. 8 (2009): 322–33.

255 **the higher brain to put things** C. A. Tassinari et al., "Biting Behavior, Aggression, and Seizures," *Epilepsia* 46, no. 5 (2005): 654–63; C. Bassetti et al., "SPECT During Sleepwalking," *The Lancet* 356, no. 9228 (2000): 484–85;

K. Schindler et al., "Hypoperfusion of Anterior Cingulate Gyrus in a Case of Paroxysmal Nocturnal Dustonia," *Neurology* 57, no. 5 (2001): 917–20; C. A. Tassinari et al., "Central Pattern Generators for a Common Semiology in Fronto-Limbic Seizures and in Parasomnias," *Neurological Sciences* 26, no. 3 (2005): 225–32.

256 **"64% of cases, with injuries in 3%"** P. T. D'Orban and C. Howard, "Violence in Sleep: Medico-Legal Issues and Two Case Reports," *Psychological Medicine* 17, no. 4 (1987): 915–25; B. Boeve, E. Olson, and M. Silber, "Rapid Eye Movement Sleep Behavior Disorder: Demographic, Clinical, and Laboratory Findings in 93 Cases," *Brain* 123, no. 2 (2000): 331–39.

256 **both the United States and the United Kingdom** John Hudson, "Common Law—Henry II and the Birth of a State," BBC, February 17, 2011; Thomas Morawetz, "Murder and Manslaughter: Degrees of Seriousness, Common Law and Statutory Law, the Model Penal Code," Law Library—American Law and Legal Information, http://law.jrank.org/pages/18652/Homicide.html.

257 **would have never consciously carried out** M. Diamond, "Criminal Responsibility of the Addiction: Conviction by Force of Habit," *Fordham Urban Law Journal* 1, no. 3 (1972); R. Broughton et al., "Homicidal Somnambulism: A Case Report," *Sleep* 17, no. 3 (1994): 253–64; R. Cartwright, "Sleepwalking Violence: A Sleep Disorder, a Legal Dilemma, and a Psychological Challenge," *American Journal of Psychiatry* 161, no. 7 (2004): 1149–58; P. Fenwick, "Automatism, Medicine, and the Law," *Psychological Medicine Monograph Supplement*, no. 17 (1990): 1–27; M. Hanson, "Toward a New Assumption in Law and Ethics," *The Humanist* 66, no. 4 (2006).

257 **attack occurred during a sleep terror** L. Smith-Spark, "How Sleepwalking Can Lead to Killing," *BBC News*, March 18, 2005.

257 **later acquitted of attempted murder** Beth Hale, "Sleepwalk Defense Clears Woman of Trying to Murder Her Mother in Bed," *Daily Mail*, June 3, 2009.

257 **sleep terrors and was found not guilty** John Robertson and Gareth Rose, "Sleepwalker Is Cleared of Raping Teenage Girl," *The Scotsman*, June 22, 2011.

257 **"Why did I do it?"** Stuart Jeffries, "Sleep Disorder: When the Lights Go Out," *The Guardian*, December 5, 2009.

259 **"his mind had no control"** Richard Smith, "Grandad Killed His Wife During a Dream," *The Mirror*, November 18, 2009.

259 **"a straight not guilty verdict"** Anthony Stone, "Nightmare Man Who Strangled His Wife in a 'Night Terror' Walks Free," *Western Mail*, November 21, 2009.

259 **you bear no responsibility** Ibid.

261 **to perfect their methods** Christina Binkley, "Casino Chain Mines Data on Its Gamblers, and Strikes Pay Dirt," *The Wall Street Journal,* November 22, 2004; Rajiv Lal, "Harrah's Entertainment, Inc.," Harvard Business School, case no. 9–604–016, June 14, 2004; K. Ahsan et al., "Harrah's Entertainment, Inc.: Real-Time CRM in a Service Supply Chain," *Harvard Business Review,* case no. GS50, May 8, 2006; V. Chang and J. Pfeffer, "Gary Loveman and Harrah's Entertainment," *Harvard Business Review,* case no. OB45, November 4, 2003; Gary Loveman, "Diamonds in the Data Mine," *Harvard Business Review,* case no. R0305H, May 1, 2003.

261 **to the cent and minute** In a statement, Caesars Entertainment wrote: "Under the terms of the settlement reached in May of 2011 between Caesars Riverboat Casino and [Bachmann], both sides (including their representatives) are precluded from discussing certain details of the case. . . . There are many specific points we would contest, but we are unable to do so at this point. You have asked several questions revolving around conversations that allegedly took place between [Bachmann] and unnamed Caesars affiliated employees. Because she did not provide names, there is no independent verification of her accounts, and we hope your reporting will reflect that, either by omitting the stories or by making it clear that they are unverified. Like most large companies in the service industry, we pay attention to our customers' purchasing decisions as a way of monitoring customer satisfaction and evaluating the effectiveness of our marketing campaigns. Like most companies, we look for ways to attract customers, and we make efforts to maintain them as loyal customers. And like most companies, when our customers change their established patterns, we try to understand why, and encourage them to return. That's no different than a hotel chain, an airline, or a dry cleaner. That's what good customer service is about. . . . Caesars Entertainment (formerly known as Harrah's Entertainment) and its affiliates have long been an industry leader in responsible gaming. We were the first gaming company to develop a written Code of Commitment that governs how we treat our guests. We were the first casino company with a national self-exclusion program that allows customers to ban themselves from all of our properties if they feel they have a problem, or for any other reason. And we are the only casino company to fund a national television advertising campaign to promote responsible gaming. We hope your writing will reflect that history, as well as the fact that none of [Bachmann's] statements you cite have been independently verified."

262 **"*did* do those nice things for me"** In a statement, Caesars Entertainment wrote: "We would never fire or penalize a host if one of their guests stopped visiting (unless it was the direct result of something the host did). And none of our hosts would be allowed to tell a guest that he or she would be fired or otherwise penalized if that guest did not visit."

264   **watch a slot machine spin around** M. Dixon and R. Habib, "Neurobehavioral Evidence for the 'Near-Miss' Effect in Pathological Gamblers," *Journal of the Experimental Analysis of Behavior* 93, no. 3 (2010): 313–28; H. Chase and L. Clark, "Gambling Severity Predicts Midbrain Response to Near-Miss Outcomes," *Journal of Neuroscience* 30, no. 18 (2010): 6180–87; L. Clark et al., "Gambling Near-Misses Enhance Motivation to Gamble and Recruit Win-Related Brain Circuitry," *Neuron* 61, no. 3 (2009): 481–90; Luke Clark, "Decision-Making During Gambling: An Integration of Cognitive and Psychobiological Approaches," *Philosophical Transactions of the Royal Society of London, Series B: Biological Sciences* 365, no. 1538 (2010): 319–30.

264   **bounced checks at a casino** H. Lesieur and S. Blume, "The South Oaks Gambling Screen (SOGS): A New Instrument for the Identification of Pathological Gamblers," *American Journal of Psychiatry* 144, no. 9 (1987): 1184–88. In a fact-checking letter, Habib wrote, "Many of our subjects were categorized as pathological gamblers based on other types of behavior that the screening form asks about. For example, it would have been sufficient for a participant to have been counted as a pathological gambler if they simply: 1) had gambled to win money that they had previously lost gambling, and 2) on some occasions they gambled more than they had intended to. We used a very low threshold to classify our subjects as pathological gamblers."

266   **circuitry involved in the habit loop** M. Potenza, V. Voon, and D. Weintraub, "Drug Insight: Impulse Control Disorders and Dopamine Therapies in Parkinson's Disease," *Nature Clinical Practice Neurology* 12, no. 3 (2007): 664–72; J. R. Cornelius et al., "Impulse Control Disorders with the Use of Dopaminergic Agents in Restless Legs Syndrome: A Case Control Study," *Sleep* 22, no. 1 (2010): 81–87.

266   **Hundreds of similar cases are pending** Ed Silverman, "Compulsive Gambler Wins Lawsuit Over Mirapex," *Pharmalot*, July 31, 2008.

266   **"gamblers are in control of their actions"** For more on the neurology of gambling, see A. J. Lawrence et al., "Problem Gamblers Share Deficits in Impulsive Decision-Making with Alcohol-Dependent Individuals," *Addiction* 104, no. 6 (2009): 1006–15; E. Cognat et al., "'Habit' Gambling Behaviour Caused by Ischemic Lesions Affecting the Cognitive Territories of the Basal Ganglia," *Journal of Neurology* 257, no. 10 (2010): 1628–32; J. Emshoff, D. Gilmore, and J. Zorland, "Veterans and Problem Gambling: A Review of the Literature," Georgia State University, February 2010, http://www2.gsu.edu/~psyjge/Rsrc/PG_IPV_Veterans.pdf; T. van Eimeren et al., "Drug-Induced Deactivation of

Inhibitory Networks Predicts Pathological Gambling in PD," *Neurology* 75, no. 19 (2010): 1711–16; L. Cottler and K. Leung, "Treatment of Pathological Gambling," *Current Opinion in Psychiatry* 22, no. 1 (2009): 69–74; M. Roca et al., "Executive Functions in Pathologic Gamblers Selected in an Ecologic Setting," *Cognitive and Behavioral Neurology* 21, no. 1 (2008): 1–4; E. D. Driver-Dunckley et al., "Gambling and Increased Sexual Desire with Dopaminergic Medications in Restless Legs Syndrome," *Clinical Neuropharmacology* 30, no. 5 (2007): 249–55; Erin Gibbs Van Brunschot, "Gambling and Risk Behaviour: A Literature Review," University of Calgary, March 2009.

268 **"they're acting without choice"** In an email, Habib clarified his thoughts on this topic: "It is a question about free will and self-control, and one that falls as much in the domain of philosophy as in cognitive neuroscience. . . . If we say that the gambling behavior in the Parkinson's patient is out of their own hands and driven by their medication, why can't we (or don't we) make the same argument in the case of the pathological gambler given that the same areas of the brain seem to be active? The only (somewhat unsatisfactory) answer that I can come up with (and one that you mention yourself) is that as a society we are more comfortable removing responsibility if there is an external agent that it can be placed upon. So, it is easy in the Parkinson's case to say that the gambling pathology resulted from the medication, but in the case of the pathological gambler, because there is no external agent influencing their behavior (well, there is—societal pressures, casino billboards, life stresses, etc.—but, nothing as pervasive as medication that a person must take), we are more reluctant to blame the addiction and prefer to put the responsibility for their pathological behavior on themselves—'they should know better and not gamble,' for example. I think as cognitive neuroscientists learn more—and 'modern' brain imaging is only about 20–25 years old as a field—perhaps some of these misguided societal beliefs (that even we cognitive neuroscientists sometimes hold) will slowly begin to change. For example, from our data, while I can comfortably conclude that there are definite differences in the brains of pathological gamblers versus non-pathological gamblers, at least when they are gambling, and I might even be able to make some claims such as the near-misses appear more win-like to the pathological gambler but more loss-like to the non-pathological gambler, I cannot state with any confidence or certainty that these differences therefore imply that the pathological gambler does not have a choice when they see a billboard advertising a local casino—that they are a slave to their urges. In the absence of hard direct evidence, I guess the best we can do is draw inferences by analogy, but there is much uncertainty associated with such comparisons."

272 **"whatever the latter may be"** William James, *Talks to Teachers on Psychology: and to Students on Some of Life's Ideals.*

272 **the Metaphysical Club** Louis Menand, *The Metaphysical Club: A Story of Ideas in America* (New York: Farrar, Straus, and Giroux, 2002).

274 **"traced by itself before"** James is quoting the French psychologist and philosopher Léon Dumont's essay "De l'habitude."

INDEX

ABOUT THE AUTHOR

CHARLES DUHIGG is an investigative reporter for *The New York Times*, where he contributes to the newspaper and the magazine. He authored or contributed to *Golden Opportunities* (2007), a series of articles that examined how companies are trying to take advantage of aging Americans, *The Reckoning* (2008), which studied the causes and outcomes of the financial crisis, and *Toxic Waters* (2009), about the worsening pollution in American waters and regulators' response.

For his work, Mr. Duhigg has received the National Academies of Sciences, National Journalism, George Polk, Gerald Loeb, and other awards, and he was part of a team of finalists for the 2009 Pulitzer Prize. He has appeared on *This American Life, The Dr. Oz Show,* NPR, *The NewsHour with Jim Lehrer,* and *Frontline.*

Mr. Duhigg is a graduate of Harvard Business School and Yale University. Before becoming a journalist, Mr. Duhigg worked in private equity and—for one terrifying day—was a bike messenger in San Francisco.

Mr. Duhigg can acquire bad habits—most notably regarding fried foods—within minutes, and lives in Brooklyn with his wife, a marine biologist, and their two sons, whose habits include waking at 5:00 A.M., flinging food at dinnertime, and smiling perfectly.

ABOUT THE TYPE

This book was set in Scala, a typeface designed by Martin Majoor in 1991. It was originally designed for a music company in the Netherlands and then was published by the international type house FSI FontShop. Its distinctive extended serifs add to the articulation of the letterforms to make it a very readable typeface.